MATHEMATICAL TECHNIQUES *in* FINANCIAL MARKET TRADING

Don K. Mak

formerly with
Federal Government Research Laboratories
Canada

MATHEMATICAL TECHNIQUES *in* FINANCIAL MARKET TRADING

NEW JERSEY • LONDON • SINGAPORE • BEIJING • SHANGHAI • HONG KONG • TAIPEI • CHENNAI

Published by

World Scientific Publishing Co. Pte. Ltd.
5 Toh Tuck Link, Singapore 596224
USA office: 27 Warren Street, Suite 401-402, Hackensack, NJ 07601
UK office: 57 Shelton Street, Covent Garden, London WC2H 9HE

Library of Congress Cataloging-in-Publication Data
Mak, Don K.
Mathematical techniques in financial market trading / Don K. Mak.
p. cm.
Includes bibliographical references and index.
ISBN-13 978-981-256-699-7 (alk. paper)
ISBN-10 981-256-699-6 (alk. paper)
1. Investments--Mathematics. 2. Finance--Mathematical models. 3. Speculation--Mathematical models. I. Title.

HG4515.3 .M35 2006
332.6401'513--dc22

2006040528

British Library Cataloguing-in-Publication Data
A catalogue record for this book is available from the British Library.

Printed in Singapore

To

my parents, whom I am indebted for my upbringing
and education,

and

my wife, whom I am thankful for her loving and care.

Preface

I finished writing the book *The Science of Financial Market Trading* in 2002. The book was written for the general public, with intended audience being the traders and investors. A number of computer programs have been included in the book for ease of application. The mathematics was kept to a minimum in the main text while the bulk of the mathematical derivations was placed in the Appendices. However, the book was actually purchased mainly by libraries and bookstores of some of the major universities and research centers around the world. It was further adopted as a textbook for a graduate course in mathematical finance by an American university.

This pleasant surprise may reflect the change in perspectives of university educators toward the trading arena for the last few years. A new discipline called "Financial Engineering" has appeared due to the demand from the financial services industry and economy as a whole. The explosive growth of computer technology and today's global financial transaction have led to a crucial demand of professionals who can quantify, appraise and predict increasingly complex financial issues. Some universities (mostly in the U.S. and Canada) are beginning to offer M.Sc. and even Ph.D. programs in financial engineering. Computing and trading laboratories are set up to simulate real life situations in the financial market. Students learn how to employ mathematical finance modeling skills to make pricing, hedging, trading, and portfolio management decisions. They are groomed for careers in securities trading, risk management, investment banking, etc.

The present book contains much more materials than the previous book. Spectrum analysis is again emphasized for the characterization of technical indicators employed by traders and investors. New indicators are created. Mathematical analysis is applied to evaluate the trading methodologies practiced by traders to execute a trade. In addition, probability theory is employed to appraise the utility of money management techniques. The book is organized in fourteen chapters.

Chapter 1 describes why the book is written. This book aims to analyze the equipment that professional traders used, and attempt to distinguish the tools from the junk.

Chapter 2 presents the latest development of scientific investigation in the financial market. A new field, called Econophysics, has cropped up. It involves the application of the principles of Physics to the study of financial markets. One of the areas concerns the development of a theoretical model to explain some of the properties of the stochastic dynamics of stock prices. There exist also growing evidences that the market is non-random, as supported by new statistical tests. In any case, market crashes have been considered to be non-random events. What the signatures are before a crash and how a crash can be forecasted will be described.

Chapter 3 analyzes the trending indicators used by traders. The trending indicators are actually low pass filters. The amplitude and phase response of one of the most popular indicators, the exponential moving average, is characterized using spectrum analysis. Other low pass filters, the Butterworth and the sinc functions are also looked into. In addition, an adaptive exponential moving average, whose parameter is a function of frequency, is introduced.

Chapter 4 modified the exponential moving average such that new designs would have less phase or time lag than the original one. It also pointed out that the "Zero-lag" exponential moving average recently designed by a trader does not live up to its claim.

Chapter 5 describes causal wavelet filters, which are actually band-pass filters with a zero phase lag at a certain frequency. The Mexican Hat Wavelet is used as an example. Calculation of the frequency where the zero phase lag occurs is shown. Furthermore, it is demonstrated how a series of causal wavelet filters with different frequency ranges can be constructed. This tool will allow the traders to monitor the long-term, mid-term and short-term market movements.

Chapter 6 introduces a trigonometric approach to find out the instantaneous frequency of a time series using four or five data points. The wave velocity and acceleration are then deduced. The method is then applied to theoretical data as well as real financial data.

Chapter 7 explains the relationship between the real and imaginary part of the frequency response function of a causal system, $H(\omega)$. Given only the phase of a system, a method is implemented to deduce $H(\omega)$. Several examples are given. The phase or time response of a system or indicator is important for a trader tracking the market movements. The method would allow them to predetermine the phase, and work backward to find out what the system is like.

Chapter 8 depicts several newly created causal high-pass filters. The filters are compared to the conventional momentum indicator currently popular with traders. Much less phase lags are achieved with the new filters.

Chapter 9 describes in detail the advantages and limitations of a new technique called skipped convolution. Skipped convolution, applied to any indicator, can alert traders of a trading opportunity earlier. However, it also generates more noise. A skipped exponential moving average would be used as an example. Furthermore, the relationship between skipped convolution and downsampled signal is illustrated.

Chapter 10 analyzes and dissects some of the popular trading tactics employed by traders, in order to differentiate the truths from the myths. It explains the meaning behind divergence of momentum (or velocity) from price. It unravels the significance of the MACD (Moving Average Convergence-Divergence) line and MACD-Histogram, but downplays the importance of the MACD-Histogram divergence.

Before putting up a trade, traders would look at charts of different timeframes to track the long-term and short-term movements of the market. The advantages and disadvantages of a long-term timeframe are pointed out in Chapter 11. This chapter also discusses how a trading plan should be put together. The popular Triple Screen Trading System is used as one of the examples.

The market is assumed to be random in Chapters 12 and 13. This modeling is good as a first approximation, and renders the application of probability theory to money management techniques practiced by traders. Chapter 12 discusses the profitability of the market at any moment in time. Chapter 13 derives and computes how traders can optimize their gain by moving the stop-loss.

The final chapter, Chapter 14, discusses the reality of financial market trading. It takes years of hard work and training to be a successful trader. In addition, the trader needs to update himself of current technology and methodology in order to keep ahead of the game.

Most of the mathematical derivations and several computer programs are listed in the Appendices.

Writing this book takes many hours of my time away from the company of my two adorable children, Angela and Anthony; and my beautiful wife, Margaret, whom I am very thankful for.

D. K. Mak
2005

Contents

Chapter 1

Introduction

Scientific theories quite often go through three stages of development: – (1) Absurdity – the idea or theory sounds so absurd that one wonders why someone would have suggested it, (2) Familiarity – there appears to be growing evidence to support the hypothesis, and people begin to familiarize themselves with the concept, and (3) Inevitability – the theory becomes so obvious in hindsight that people would think why it was not recognized earlier and why it has taken so long for the community to come to accept it.

Is the financial market not random? Fifty years ago, the academia would think it was ridiculous to say that the market was non-random. Since then, there have appeared journal papers challenging the random walk theory. At the moment, some academics would conclude that the market is non-random (see details in Chapter 2). However, the debate is still on, and there could be many years before the final verdict is in.

During all these time, the market traders could not care less what the academics think. They swear, by their own observation and experience, that the market is not random. Some even claim even if it were random, with good money management, they can still make a profit from the market. They facilitate their own methods to trade. Some do consistently make money from the market year after year. They design indicators to forecast which way the market is heading. And they devise trading systems to enter and exit the market. However, no trader seems to care to analyze their indicators and methodologies mathematically, nor do they try to characterize them. Their tools range from the very useful to complete garbage.

This scenario is somewhat similar to alternative medicine thirty years ago. Then, alternative medicine was unconventional, unproven, and unorthodox, and was ignored by the mainstay medical researchers. However, some of the alternate approaches do represent many years of experience of the practitioners by trial and error, and can contain some truths. They may even depict innovative means to problems conventional medicine has no cure. But, then, of course, some of the alternative medicine is eccentric and harmful. It was fortunate that medical researchers did finally take a serious note at these alternative therapies, and apply scientific methods to study them. It would be up to them to differentiate the grass from the weeds.

The tools employed by the market traders have a similar script. Some professional traders, by trial and errors, pick certain indicators as their arsenals, and make consistent profits from the market, even though they do not exactly understand the properties of their accouterments. Other traders advertise their indicators, and black box methodologies, and claim they can perform miracles. Believers wind up losing their shirts in the market.

It is the purpose of this book to analyze their tools mathematically, and display their characteristics. Spectrum analysis is emphasized. Some of the ideas have been presented earlier [Mak 2003]. We will expand on those ideas. We will point out why some of the traders' techniques work, and why some do not. In addition, we will also look at how a good trading plan can be put together, and how, according to probability theory, some of the money management techniques employed by traders do make profitable sense. Furthermore, we will invent some new indicators, which have less time or phase lag than the ones currently used by traders. These would allow them to pick up market signals earlier. We hope that this presentation will be useful to the trading community.

Chapter 2

Scientific Review of the Financial Market

How the financial market has been modeled in different endeavors has been described by Mak [2003]. From all perspectives, it seems as if it would be best modeled as a complex phenomenon. A complex system contains a number of agents who are intelligent and adaptive. The agents make decisions on the basis of certain rules. They can modify old rules or create new rules as new information arises. They know at most what a few other agents are doing. They then decide what to do next based upon this limited information [Waldrop 1992, Casti 1995, Johnson et al 2003]. Scientists and mathematicians have been trying to draw some conclusions from the complex financial system. Some of their recent attempts are described below.

2.1 Econophysics

Over the past two decades, a growing number of physicists has become involved in the analysis of the financial markets and economic systems. Using tools developed in statistical mechanics, they were able to contribute to the modelling of the dynamics of the economy in a practical fashion. A new field, known as econophysics, has thus emerged [Mantegna and Stanley 2000]. The field benefits from the large database of economic transactions already recorded. Several findings are described below.

2.1.1 *Log-Normal Distribution of Stock Market Data*

In 1900, Bachelier wrote that price change in the stock market followed a one-dimensional Brownian motion, which has a normal (Gaussian) distribution [Mandelbrot 1983, 1997]. Since the 1950's, the distribution of the stock price changes has been considered by several mathematicians. The Gaussian distribution was soon replaced by the log-normal distribution. Stock prices are performing a geometric Brownian motion, and the differences of the logarithms of prices are Gausssian distributed. A full review of these investigations can be found in Crow and Shimizu [1988].

Recently, Antoniou et al [2003] analyzed the statistical relations between prices and corresponding traded volumes of a number of stocks in the United States and European markets. They found that, for most stocks, the statistical distribution of the daily closing prices normalized by corresponding traded volumes (price/volume) fits well the log-normal function. The statistical distribution is given by:

$$f(x) = \frac{A}{\sqrt{2\pi}\sigma x} \exp^{-(1/2\sigma^2)(\ln x - \mu)^2} \qquad (2.1)$$

where x = price/volume,
A is a normalizing factor,
σ is the dispersion,
μ is the mean value.

For some other stocks, the log-normal function is attained after application of a detrending process.

They have also discovered that the distributions of the stocks' traded volumes normalized by their trends fit closely the log-normal functions. However, market indices have significantly more complicated characters, and cannot be approximated by log-normal functions.

Other stock market models have been proposed by other researchers. They are particularly employed to explain the observation that the tails of distributions in real data are fatter than expected for a log-normal distribution.

2.1.2 *Levy Distribution*

Among the alternative models proposed is the conjecture that price change is governed by a Levy stable distribution [Mandelbrot 1983, 1997]. The distribution is leptokurtic, i.e., it has wings larger than those of a normal process. It has described well the price variations of many commodity prices, interest rates and stock market prices [Mandelbrot 1983].

In 1995, Mantegna and Stanley showed that the central part of the probability distribution of the Standard & Poor 500 index (S & P 500) can be described by the Levy stable process. Furthermore, when the process is rescaled, the transformations fit well time intervals spanning three orders of magnitude, from 1,000 min to 1 min. The Levy distribution will be described in more detail in Chapter 12.

2.1.3 *Tsallis Entropy*

Time evolving financial markets can be described in terms of anomalously diffusing systems, where a mean-square displacement scales with time, t, according to a power-law, t^{α} [Michael and Johnson, 2002]. Anomalously diffusion systems can be treated by employing the nonlinear Fokker-Planck equation associated with the Ito-Langevin process [Tsallis and Bukman 1996]. The solution of the equation is a time-dependent probability distribution which maximize the Tsallis entropy. The probability distribution can be written as :

$$P(x,t) = [1/Z(t)]\{1 + \beta(t)(q-1)[x - x^*(t)]^2\}^{-1/(q-1)} \tag{2.2}$$

where x is a price change during a time interval t,
x^* is the mean,
Z (a normalization constant) and β are Lagrange multipliers,
q is a Tsallis parameter.

The 1-min-interval data of the S & P 500 stock market index collected from July 2000 to January 2001 has been used as a test case. A nonlinear χ^2 fit of Eq (2.2) for t = 1 minute yields q = 1.64 +/- 0.02, β = 4.90 +/- 0.11. Z can be calculated to be 1.09 +/- 0.02. The data fits the

the probability distribution, P, quite well. P, using the above parameters, are plotted in Fig 2.1.

It can be shown that, in compliance with probability theory :

$$\int P(x,t) \approx 1 \tag{2.3}$$

The data were then fitted to different time intervals t, viz, 10 min and 60 min with q = 1.64 fixed, and β determined by the fit. These data again fit the Tsallis distribution, P, quite well. This shows that P yields a solution to the time evolving Fokker-Planck equation, which describes an anomalously diffusing system. Anomalus diffusion implies that price changes during successive time intervals are not indpendent. This is consistent with traders responding to earlier price changes. The diffusion of the financial market indicates correlation, and hence a non-trivial time dependence.

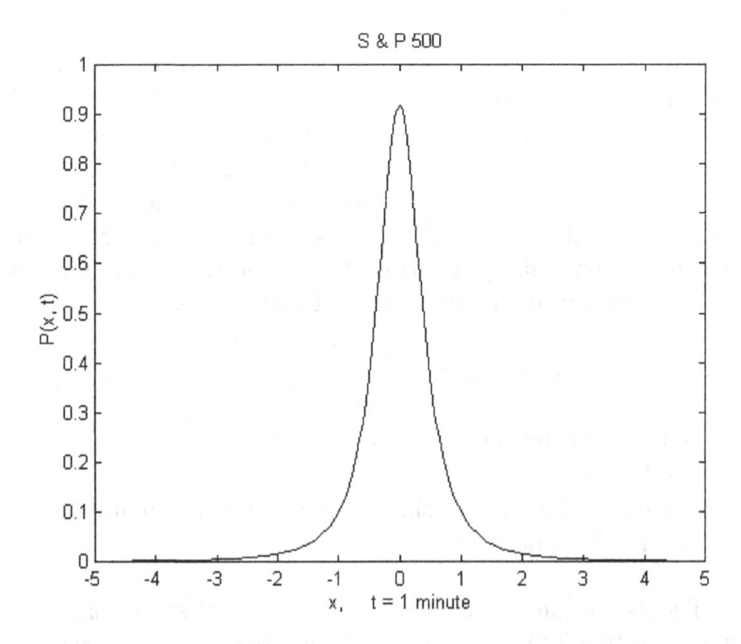

Fig 2.1 Tsallis distribution, P, of the 1-min-interval data of the S & P 500 index.

Symmetrical probability distributions of market price changes can imply that the market is random. To a first approximation, it probably is. However, looking at the market in more detail, is it really random? We will take a look at this issue in the next section.

2.2 Non-Randomness of the Market

The academia has been insisting that the market can be described by the Random Walk Hypothesis. The randomness is achieved through the active participation of many investors and traders. They aggressively digest any information that is available, and incorporate those information into the market prices, thus eliminating any profit opportunities. Therefore, in an informationally efficient market, price changes must be unforecastable. The Efficient Market Hypothesis actually states that, in an active market that includes many well-informed rational investors, securities will be appropriately priced and reflect all available information. The Efficient Market Hypothesis is considered as a close relative of the Random Walk Hypothesis.

2.2.1 *Random Walk Hypothesis and Efficient Market Hypothesis*

In the last decade or so, some academics are having a second thought about the randomness of the market. In the book "A Non-Random Walk Down Wall Street", Lo and MacKinlay [1999] has demonstrated convincingly that the financial markets are predictable to some degree.

They first pointed out that the Random Walk Hypothesis and the Efficient Markets Hypothesis are not equivalent statements. One does not imply the other, and vice versa. In other words, random prices does not imply a financial market with rational investors, and non-random prices does not imply the opposite. The Efficient Market Hypothesis only takes into account the rationality of the investors and information available, but not of the risk that some investors are willing to take. If a security's expected price change is positive, an investor may choose to hold the asset and bear the associated risk. On the contrary, if the investor wants to avert risk at a certain time, he may choose to dump his security to avoid having unforecastable returns. At any time, there are always investors who have unexpected liquidity needs. They would trade and possibly lose money. This does not mean that they do not know the information, nor are they irrational.

Lo and MacKinlay [1999] further employed a variance-ratio test (see section 2.2.2) to show that the market was not random. They also discovered that they were not the first study to reject the random walk. Papers describing the departures from random walk have been published since 1960, but were largely ignored by the academic community. They then concluded that the apparent inconsistency of their findings and the general support of the Random Walk Hypothesis is largely caused by the misconception that the Random Walk Hypothesis is equivalent to the Efficient Market Hypothesis, and the dedication of the economists to the latter.

2.2.2 *Variance-Ratio Test*

Lo and MacKinlay [1999] has proposed a test for the random walk based on the comparison of variances at different sampling intervals, as variance is considered a more sensitive parameter than a mean when data is sampled at finer intervals. The test makes use of the fact that the variance of the increments of a random walk is linear with respect to the sampling interval. If stock prices are induced by a random walk, then, the variance of a monthly sample must be four times as large as that of a weekly sample.

They employed for computation the 1216 weekly observations from September 6, 1962 to December 26, 1985 of the equal-weighted Center for Research in Security Prices (CRSP) returns index. The modified variance ratios of the 2-week, 4-week, 8-week and 16-week returns to the 1-week return were calculated. All these ratios are statistically different from 1 at the 5% level of significiance. This can be compared with the random walk where the modified variance ratio is 1. Thus, they concluded that the random walk null hypothesis could be rejected. They further pointed out that the modified variance ratio of 2-week return to 1-week return should be approximately equal to 1 plus the first-order autocorrelation coefficient estimator of weekly returns. The first-order autocorrelation for weekly returns thus calculated is approximately 30%. Therefore, the random walk null hypothesis can be easily rejected even on the basis of autocorrelation alone. Using the same variance analysis with daily returns, they also found that the case against the random walk was equally compelling.

They then changed the base observation period to 4 weeks. The modified variance ratios of the 8-week, 16-week, 32-week and 64-week returns to the 4-week return were calculated. The ratios showed that the random walk model could not be rejected. The result is consistent with previous studies which have also found weak evidence against the random walk when using monthly data.

All these results are further supported by a modified R/S statistic test which will be described in the next section.

2.2.3 *Long-Range Dependence?*

There are many theories that business cycles exist, and economics time series can exhibit long-range (monthly and yearly) dependence. To test this dependence, Lo and MacKinlay [1999] modified a "range over standard deviation" ("R/S") statistic which was first proposed by the English hydrologist Harold Edwin Hurst and later refined by Mandelbrot. The R/S statistic is the range of partial sums of deviations in a time series from its mean, rescaled by its standard deviation. However, it cannot distinguish between short-range and long-range dependence. The R/S statistic has to be modified so that its statistical behavior is invariant over short-term memory, but deviates over long-term memory. The modified statistic was then applied to daily and monthly CRSP stock return indexes over several sample periods. After correcting for short-range dependence, there was no evidence that long-range dependence existed. The test showed that there was little dependence in daily stock returns beyond one or two months.

Furthermore, the autocorrelograms of the daily and monthly stock return indexes were also plotted, with a maximum lag of 360 for daily returns, and 12 for monthly. It was found that for both indexes, only the lowest order autocorrelation coefficients were statistically significant.

Thus, the long-range dependence of stock returns uncovered by previous studies may not be the long-term memory in the time series, but simply the result of short-range dependence.

2.2.4 *Varying Non-Randomness*

In an update to their original variance ratio test for the weekly US stock market indexes, Lo and MacKinlay [1999] found that the more current data (1986 – 1996) conformed more closely to the random walk than the original 1962 – 1985 data. Upon investigation, they discovered that over the past decade, a few investment firms had exercised daily equity trading strategies devised to exploit the kind of patterns they revealed in 1988. This can provide a plausible explanation why recent data is more random. This observation also supports the idea that the market is a complex system. Traders, being very adaptive, will learn new information, and actively modify their rules to their advantages [Mak 2003]. This, in turn, will affect the market, and narrow any profitable opportunities.

2.3 Financial Market Crash

While probability distributions, like the Levy distribution, describes the central part of the distribution of market price variation quite well, they do not match the rare events, like the market crashes. Market crashes are outliers. Outliers are extreme values that do not fit the model. If so, another model needs to be considered to explain these rare occurrences.

2.3.1 *Log-Periodicity Phenomenological Model*

Sornette [2003] first formed an hypothesis that the time evolution of market prices were random walks. Using this hypothesis, he derived a result where the distribution of market drops would be exponential. Comparing this result with that constructed from indices of various countries, he found an apparent discrepancy, especially with respect to the large market drops usually known as crashes. It was then concluded that crashes could not be completely random. If so, they might be somewhat forcastable as other catastrophes, like earthquakes and ruptures of pressure tanks.

Sornette [2003] drew comparison of crashes to critical phenomena and nonlinear interactions in modern physics. He proposed a signature before a crash, a "bubble", as a log-periodic correction imposed on a power law for an observable exhibiting a singularity at time t_c, where t_c is the time where the crash has the highest probability to occur. The oscillatory market index data is fitted to the following mathematical expression:

$$F_{1p}(t) = A_2 + B_2(t_c - t)^m [1 + C \cos(\omega \log((t_c - t)/T))] \tag{2.4}$$

The power law, $A_2 + B_2(t_c - t)^m$, represents the advancing price in the bull market. The price accelerates and eventually ends in a spike. This corresponds to a pattern described as a "half moon" by technical analysts [Prechter and Frost 1990]. Sornette noted the presence of oscillatory-like deviations in the trend. The oscillation is described by the cosine function of the logarithm of $(t_c - t)/T$. A_2, B_2, t_c, m, C, ω and T are all fitting parameters. These parameters, of course, vary for different bubbles.

It should be noted that, unlike some catastrophes like earthquakes, bubbles and crashes are events occurred in financial markets, which are complex systems. A complex system contains a number of agents, who are intellegent and adaptive [Waldrop 1992; Casti 1995; Mak 2003]. They make decisions and behave according to certain rules. They can change the rules as new information arises. Thus, phenomena of natural disasters may be quite different from rare events in the financial markets.

Furthermore, critical phenomena and phase transitions in thermodynamics and statistical mechanics are interesting and significant areas to be studied [Stanley 1971; Huang 1963]. Nevertheless, comparing market crashes to these critical events can only be qualitative. In the market, rules can be changed. For example, following the market crash of October, 1987, the U. S. Securities and Exchange Commission installed the so-called circuit breakers to head off one-day stock market tumbles in the future. The market will be halted after a one-day decline of 10% in the Dow Jones Industrial Average. This inactive period will allow the traders to pause and evaluate their positions. These circuit breakers will definitely affect market crashes as well as their precursory patterns in the future.

2.3.2 *Omori Law*

The relaxation dynamics of a financial market just after a crash can be viewed in terms of a complicated system when the system experiences an extreme event. The relaxation is described by a power-law distribution, which implies that rare events can occur with a finite non-negligible probability. It has been shown that the dynamics follow the Omori Law [Lillo and Mantegna 2003]. The law describes the nonstationary period observed after a big earthquake. It says that, after a main earthquake, the number of aftershock earthquakes per unit time measured at time t, n(t), decays as a power law. The law is written as

$$n(t) = K(t + \tau)^{-p} \qquad (2.5)$$

where K and τ are two positive constants, and p is the exponent. The cumulative number of aftershocks, N(t), observed until time t after the earthquake can be obtained by integrating Eq (2.5) between 0 and t. N(t) is thus given by

$$N(t) = K[(t+\tau)^{1-p} - \tau^{1-p}]/(1-p) \qquad p \neq 1 \qquad (2.6a)$$

$$= K \ln(t/\tau + 1) \qquad p = 1 \qquad (2.6b)$$

When the 1-min logarithm changes of the S & P 500 index, r(t), (a quantity essentially equivalent to index return), is investigated after a financial crash, it has been found that the number of times |r(t)| exceeds a given threshold, behaves like the Omori Law - somewhat similar to n(t). While the value of the exponent p for earthquakes ranges between 0.9 and 1.5, p for the financial market varies in the interval between 0.70 and 0.99. It has been further noted that the index return cannot be modeled in terms of independent identically distributed random process after a market crash. This observation would substantiate the claim that the market is not a random phenonmenon.

Chapter 3

Causal Low Pass Filters

Trending indicators are used by traders to identify trends. They basically smooth the input data [Mak 2003]. They are actually low pass filters which filter off the high frequencies, leaving the low frequencies behind.

A filter is said to be causal if the output of the filter depends only on present and past inputs, but does not depend on future inputs [Proakis and Manolakis 1996, Strang and Nguyen 1997]. For traders, the indicators have to be causal as no future data is available.

3.1 Ideal Causal Trending Indicator

An ideal causal trending indicator to the traders would look like a brick wall filter whose bandwidth ranges from 0 to a cutoff frequency ω_c. (ω, in units of radians, is quite often called the circular frequency, and is equal to $2\pi f$, where f is the reciprocal of the period, T). Frequencies larger than ω_c would be eliminated, while frequencies larger than 0 and less than ω_c will be kept with amplitude unchanged. In addition, for those frequencies kept, the phase would be unchanged, i.e., there is no time or phase lag. However, this design is mathematically impossible.

We have to live with something less ideal. One of the causal trending indicators favored by traders is the exponential moving average [Pring 1991, Elder 1993, 2002, Mak 2003]. This indicator will be described in the next section.

3.2 Exponential Moving Average

An exponential moving average (EMA) is a better tool than a simple moving average (SMA). A simple moving average takes the average of the input data with equal weights [Elder 1993, Mak 2003]. An exponential moving average gives greater weight to the latest data and thus responds to changes faster. It does not drop old data suddenly the way an SMA does. Old data fades away.

The equation for the output response of an EMA is given by

$$y(n) = \alpha x(n) + (1-\alpha)y(n-1) \quad (3.1)$$

$$\text{where } \alpha = 2/(M+1) \quad (3.2)$$

M is a positive integer chosen by the trader and is often called the length of the EMA. Thus, α has to be equal or less than 1.

Equation (3.1) makes use of an output response that has already been processed. Filters that employ previously processed values are sometimes called recursive filter. To calculate the frequency response of EMA, the z-transform of Eq (3.1) is taken [Broesch 1997, Proakis and Manolakis 1996].

$$Y(z) = \alpha X(z) + (1-\alpha)z^{-1}Y(z) \quad (3.3)$$

where $z = r\exp(i\omega)$ is a complex number in the complex plane, r being the magnitude of z. Y(z) is the transform of the output and X(z) is the transform of the input.

Defining the transfer function as the output of the filter over the input of the filter

$$H(z) = Y(z)/X(z) \quad (3.4)$$

we get, for EMA

$$H(z) = \frac{\alpha}{1-(1-\alpha)z^{-1}} \tag{3.5}$$

The EMA has a single pole in its transfer function. A pole is a zero of the denominator polynomial of the transfer function H(z). Restricting z in the complex plane to exp(iω) on the unit circle (i.e. r = 1), the frequency response function H(ω) is given by

$$H(\omega) = \frac{\alpha}{1-(1-\alpha)\exp(-i\omega)} \tag{3.6}$$

The magnitude of H(ω) is given by [Lyons 1997]

$$|H(\omega)| = \frac{\alpha}{[1-2(1-\alpha)\cos\omega + (1-\alpha)^2]^{1/2}} \tag{3.7}$$

The phase is given by

$$\phi(\omega) = \tan^{-1}\left[\frac{-(1-\alpha)\sin\omega}{1-(1-\alpha)\cos\omega}\right] \tag{3.8}$$

The magnitude and phase of H(ω) of EMA are plotted in Fig 3.1(a) and (b) respectively for M = 3 and M = 6, from $\omega = 0$ to π.

Traders quite often like to express the phase lag in terms of a lag in the number of data points (bars) [Ehlers 2001]. The lag in the number of data points can easily be calculated by dividing the phase, ϕ, by the circular frequency, ω. Fig 3.1(b) can be re-plotted in Fig 3.1(c) in terms of the lag in the number of bars. It should be noted that for M = 3, the phase lag is less than 1 bar. This small lag makes the EMA a rather popular tool for traders.

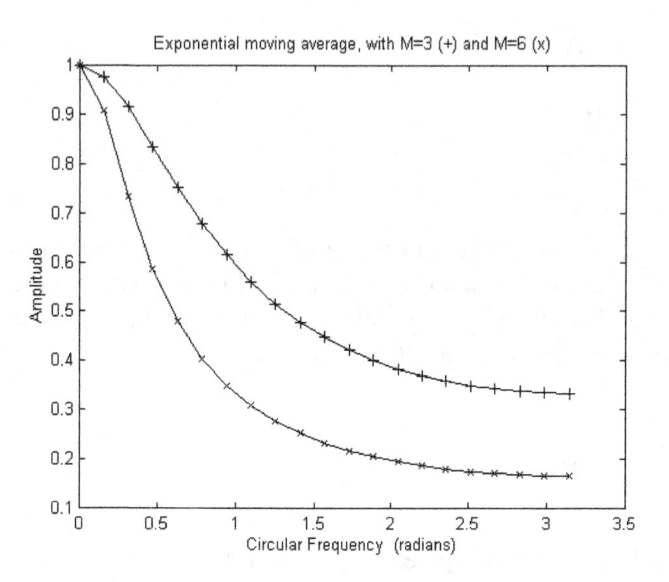

Fig 3.1(a) Amplitude response of an exponential moving average with M = 3 (marked as +) and M = 6 (marked as x) is plotted versus circular frequency ω from 0 to π.

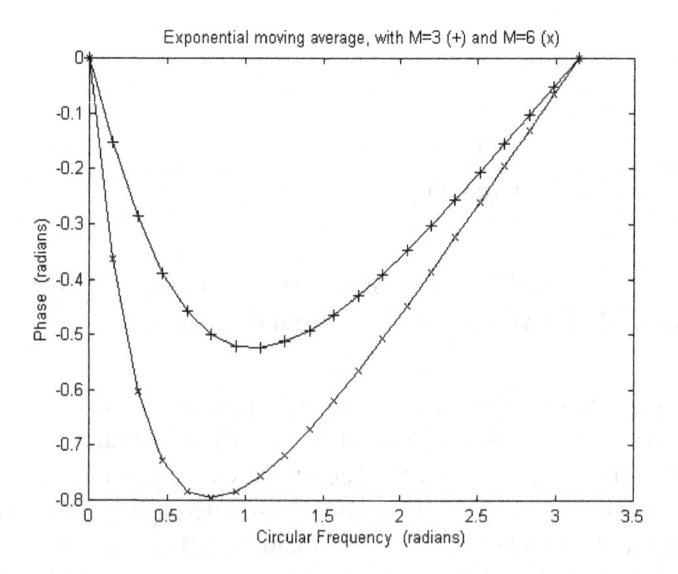

Fig 3.1(b) Phase response of an exponential moving average with M = 3 (marked as +) and M = 6 (marked as x).

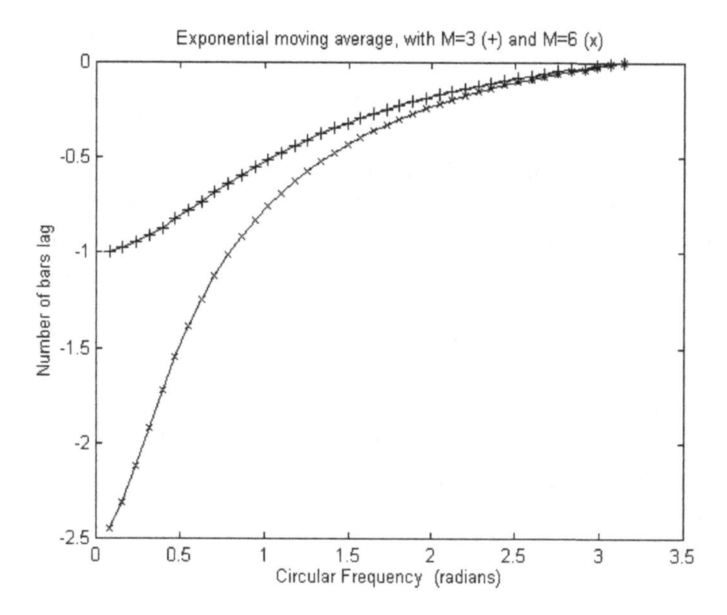

Fig 3.1(c) Phase response in terms of the number of bars lag of an exponential moving average with M = 3 (marked as +) and M = 6 (marked as x).

3.3 Butterworth Filters

One of the most commonly used low pass filters among electrical engineers is the Butterworth filter [Hamming 1989, Oppenheim et al 1999, Hayes 1999, Ehlers 2001]. We will analyze this filter for comparison purpose. The system function, H(s) of the Butterworth filter can be given by [Hayes 1999]:

$$H(s) = \frac{1}{s^N + a_1 s^{N-1} + \ldots.. + a_{N-1} s + a_N} \tag{3.9}$$

where

$s = j\omega/\omega_c$

ω_c is the 3-db cutoff circular frequency

N is the order of the filter (number of poles in the transfer function)

a's are coefficients of the polynomial and can be found in Hayes [1999].

The amplitude and phase of Eq (3.9) for N = 1, 2, 3, 4 with $\omega_c = 1$ are plotted in Fig 3.2(a) and 3.2(b) respectively. From Fig 3.2(a), it can be seen that the amplitude response of the filter decreases monotonically with ω. As the filter order N increases, the transition band, the region between the passband, where signals are passed, and the stopband, where signals are filtered off, becomes narrower. From Fig 3.2(b), we can see that a single pole Butterworth filter has a much larger phase lag than the single pole exponential moving average. As the number of poles increase, the phase lag gets larger. Thus, while the Butterworth filter is a very useful filter for electrical engineers, it is not so useful to traders.

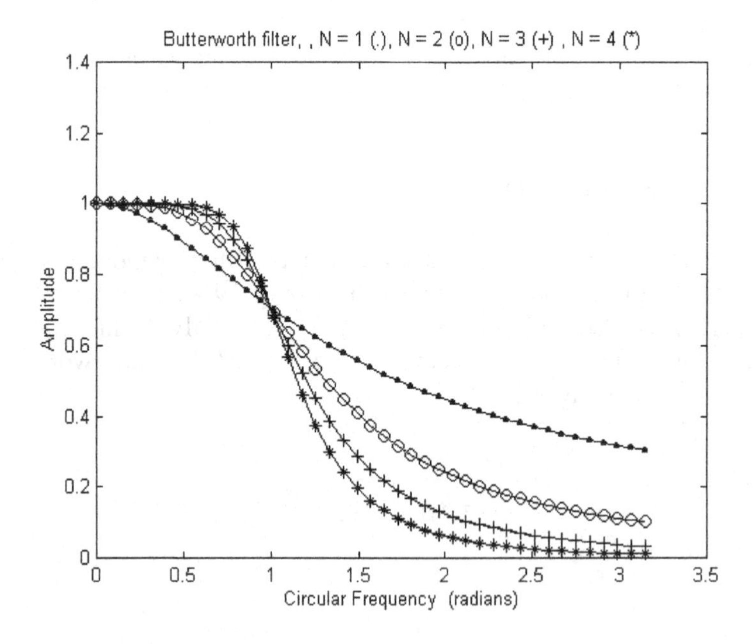

Fig 3.2(a) Amplitude response of the Butterworth filter for N = 1, 2, 3, 4 with $\omega_c = 1$.

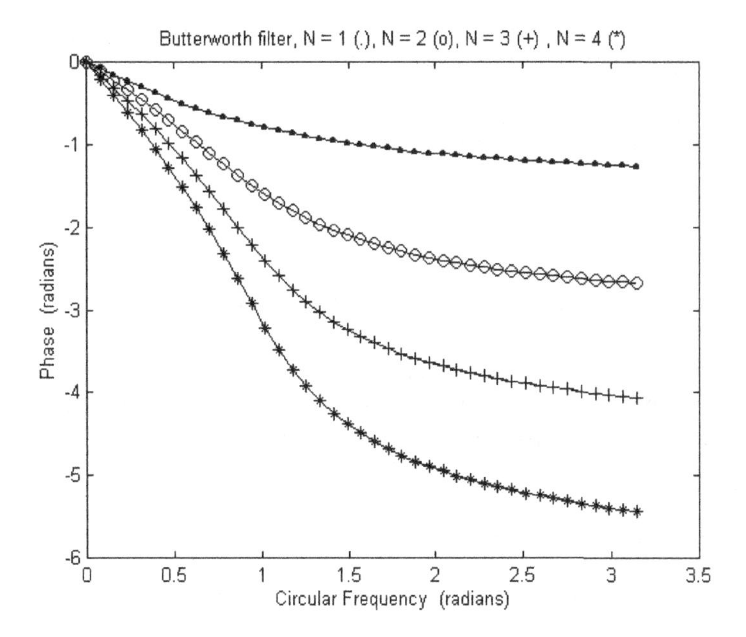

Fig 3.2(b) Phase response of the Butterworth filter for N = 1, 2, 3, 4 with $\omega_c = 1$.

3.4 Sinc Function, n = 2

Sinc functions have been mentioned in Mak [2003]. They can be considered as scaling functions, which are the father of wavelets [Hubbard 1998]. Scaling functions are actually low-pass filters while wavelets are band-pass filters. Wavelets will be considered in more detail in Chapter 5. Here we discuss the sinc functions, which can be regarded as ideal low-pass filters [Strang 1997]. A sinc function is considered a very good low pass filter as the frequency response looks like a step function with a cutoff frequency, ω_c, eliminating signals with frequencies above the cutoff. It looks like a brick wall filter.

The discrete sinc function can be written as:

$$h_n(k) = \frac{\sin\frac{\pi k}{n}}{\pi k} \qquad \text{where} \quad n = 1,2,3,.... \tag{3.10}$$

Eqn (3.10) can be considered as the scaling function for the sinc wavelets [Mak 2003, P211]. For n = 2, the discrete since function is written as

$$h_2(k) = \frac{\sin \frac{\pi k}{2}}{\pi k} \tag{3.11}$$

The coefficients, $h_2(k)$ is the unit impulse response of a low pass filter. They are plotted in Fig 3.3 for k = 0,1, ...120. For k larger than 120, $h_2(k)$ is approximately equal to zero, and does not have a large impact on the moving average of the data that it is convoluting. The coefficients $h_2(k)$ are listed in Appendix 1.

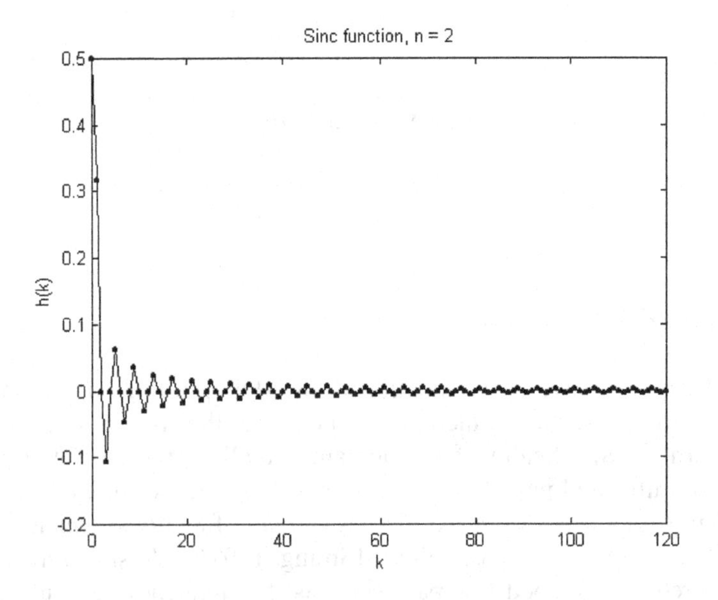

Fig 3.3 The coefficients, $h_2(k)$, of the sinc function with n = 2.

The Fourier Transform of $h_2(k)$ can provide the frequency characteristics of the low pass filter. The amplitude and phase of the Fourier Transform are plotted in Fig 3.4(a) and (b) versus circular frequency ω. Fig 3.4(a) shows that the low pass filter has a cutoff

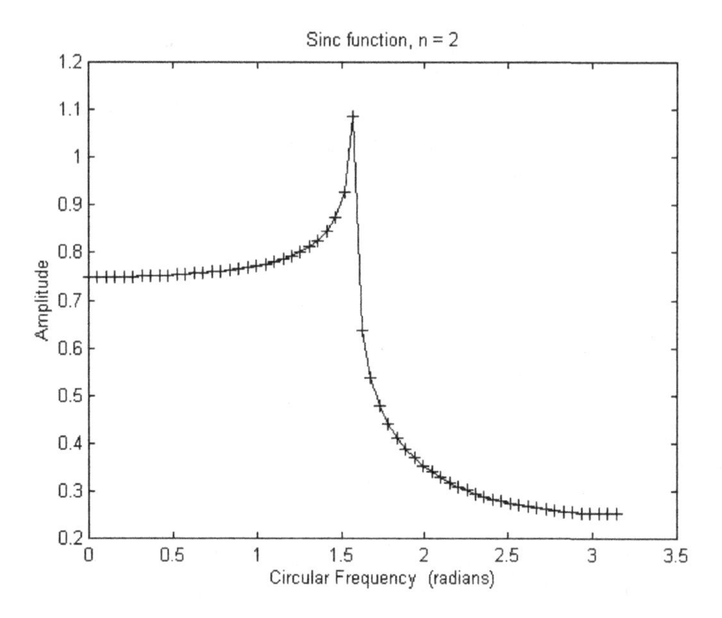

Fig 3.4(a) Amplitude response of the sinc function with n = 2.

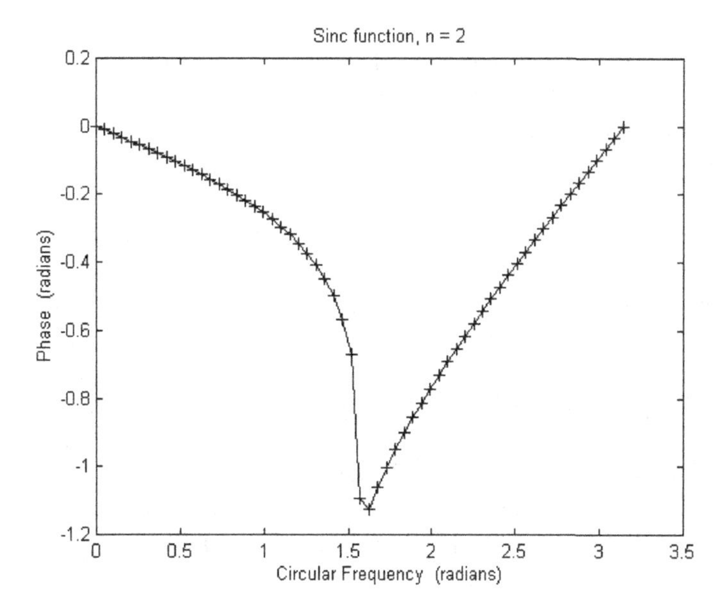

Fig 3.4(b) Phase response of the sinc function with n = 2.

frequency at $\pi/2$. Fig 3.4(b) shows that it has a phase lag of less than 0.41 radians for circular frequency less than 1.3 radians. This means that it has less phase lag than the exponential moving average with M = 3 for this frequency range. However, for frequencies close to $\pi/2$, the phase lag increases drastically.

3.5 Sinc Function, n = 4

For n = 4, the discrete sinc function is written as

$$h_4(k) = \frac{\sin\frac{\pi k}{4}}{\pi k} \tag{3.12}$$

The coefficients, $h_4(k)$ is the unit impulse response of a low pass filter. They are plotted in Fig 3.5 for k = 0,1, ...120. For k larger than 120, $h_4(k)$ is approximately equal to zero, and does not have a large impact on the moving average of the data that it is convoluting. The coefficients $h_4(k)$ are listed in Appendix 1.

The Fourier Transform of $h_4(k)$ can provide the frequency characteristics of the low pass filter. The amplitude and phase of the Fourier Transform are plotted in Fig 3.6(a) and (b) versus the circular frequency ω. Fig 3.6(a) shows that the low pass filter has a cut-off frequency at $\pi/4$. Fig 3.6(b) shows that it has a phase lag of less than 0.4 radians for circular frequency less than 0.5 radians. This means that it has less phase lag than the exponential moving average with M = 3 for this frequency range. However, for frequencies close to $\pi/4$, the phase lag increases drastically.

Despite some of its shortcomings, the sinc functions can be useful low-pass filters or trending indicators for traders due to their brick wall nature and small phase lag for part of the frequency range. They can be particularly useful when the sinc wavelet filters are used at the same time [Mak 2003]. Their potentiality should be exploited.

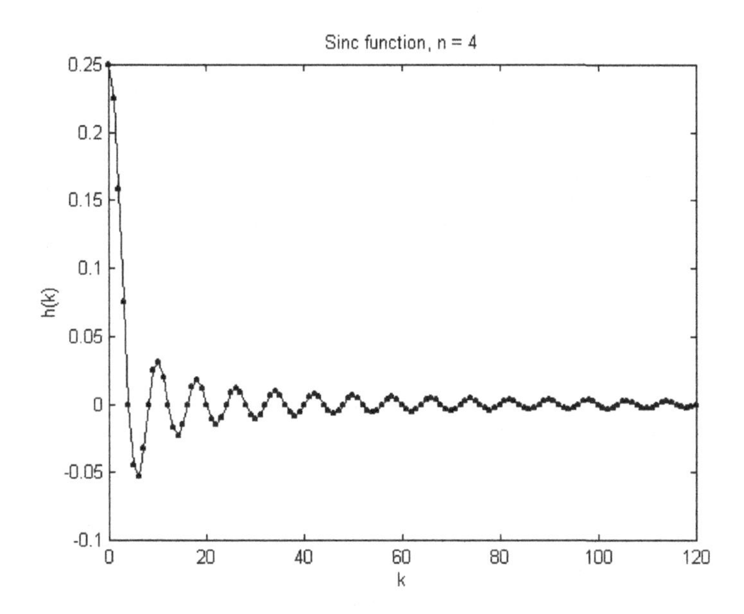

Fig 3.5 The coefficients, $h_4(k)$, of the sinc function with n = 4.

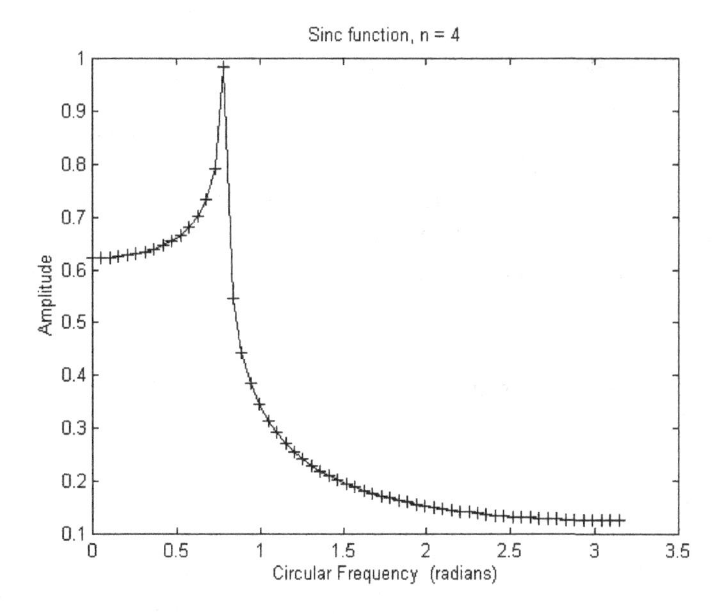

Fig 3.6(a) Amplitude response of the sinc function with n = 4.

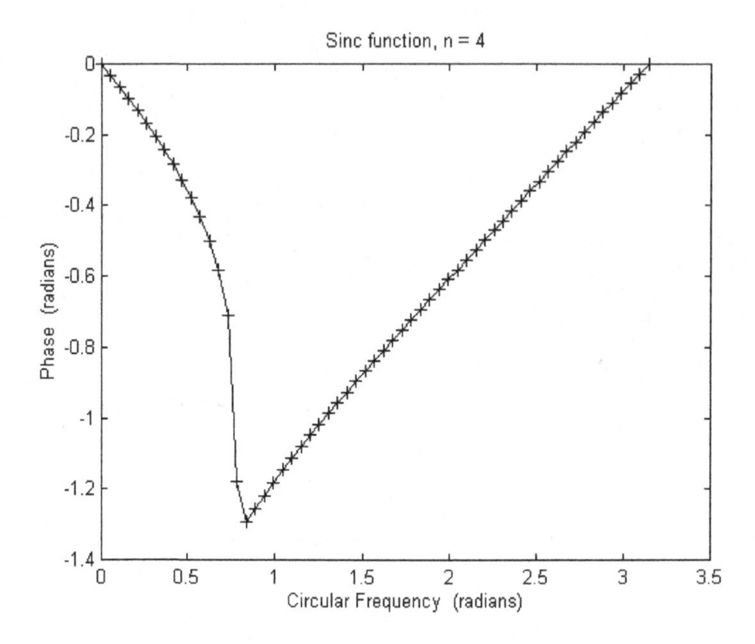

Fig 3.6(b) Phase response of the sinc function with n = 4.

3.6 Adaptive Exponential Moving Average

Moving averages that will adapt to the market environment have been suggested [Ehlers 2001]. One possibility is to vary the parameter, α, in the original exponential moving average. Several illustrations have been given by Ehlers [2001]. Here, we create one example where α is a function of the circular frequency ω. The estimation of ω will be considered in Chapter 6.

The equation for the output response of an EMA has been given by (Eq (3.1) and (3.2)) :

$$y(n) = \alpha x(n) + (1-\alpha)y(n-1) \tag{3.13}$$

where $\alpha = 2/(M+1)$ (3.14)

To make the EMA adaptive, α would be made a variable ranging between a maximum c (< 1) when ω is less than or equal to ω_0 and a minimum value when ω is equal to π. The maximum and minimum values will be chosen by the trader. Thus, α would be dependent upon the circular frequency ω, and can be written as follows:

$$\alpha = \begin{cases} c & \omega \leq \omega_0 \\ a/\omega + b & \omega_0 < \omega \leq \pi \end{cases} \tag{3.15}$$

Thus, for large ω, α will be smaller and M would be larger. This simply means that noisier data would be smoothed with an EMA with a larger M. We will set the maximum value of α to be 0.5 and the minimum value of α to be 0.05. When ω is less than or equal to 1 radian, α will be set to 0.5. When ω is equal to π radian, α will be set to 0.05. Thus, substituting (1, 0.5) and (π, 0.05) into Eqn (3.15) will yield:

$$\alpha = \begin{cases} 0.5 & \omega \leq 1 \\ 0.66/\omega - 0.16 & 1 < \omega \leq \pi \end{cases} \tag{3.16}$$

The amplitude and phase of the transfer function H(z) for $\alpha = 0.5$ (i.e. M = 3), $\alpha = 0.05$ (M = 39) and the adaptive α is plotted in Fig 3.7(a) and 3.7(b) respectively. It can be noted that the phase of the adaptive EMA is always less than 0.7 radians. The number of bars lag for different α's are plotted in Fig 3.7(c) and expanded in Fig 3.7(d) for $\alpha = 0.5$ (i.e. M = 3) and the adaptive α, showing that the adaptive EMA has at most 1 bar lag. Thus, this adaptive EMA is a much better trending tool than the original EMA. However, it has the disadvantage that the circular frequency, ω, has to be estimated accurately.

In the next chapter, we will see how the lag of an EMA can be reduced by some other means.

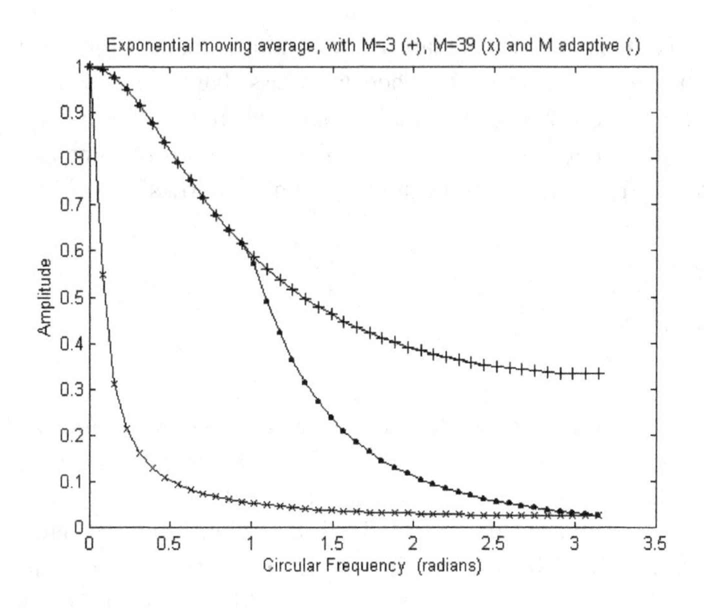

Fig 3.7(a) Amplitude response of the exponential moving average with M = 3 (plotted as +), M = 39 (plotted as x), and an adaptive M (plotted as .) .

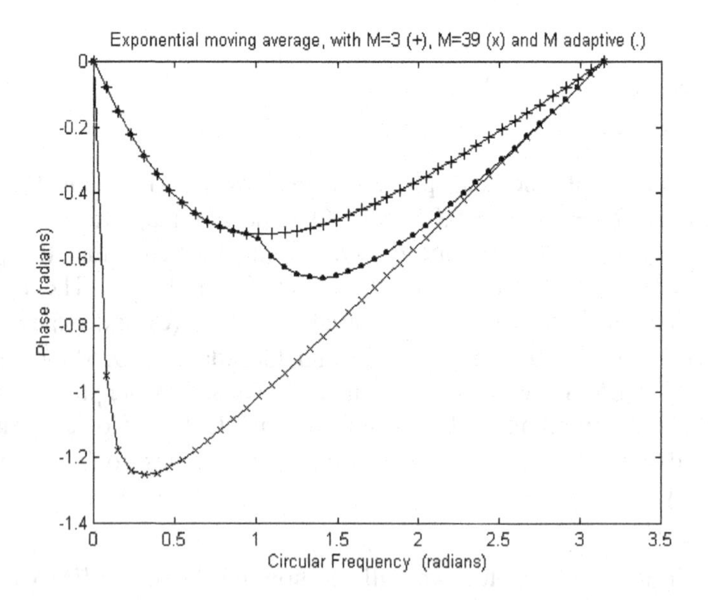

Fig 3.7(b) Phase response of the exponential moving average with M = 3 (plotted as +), M = 39 (plotted as x), and an adaptive M (plotted as .) .

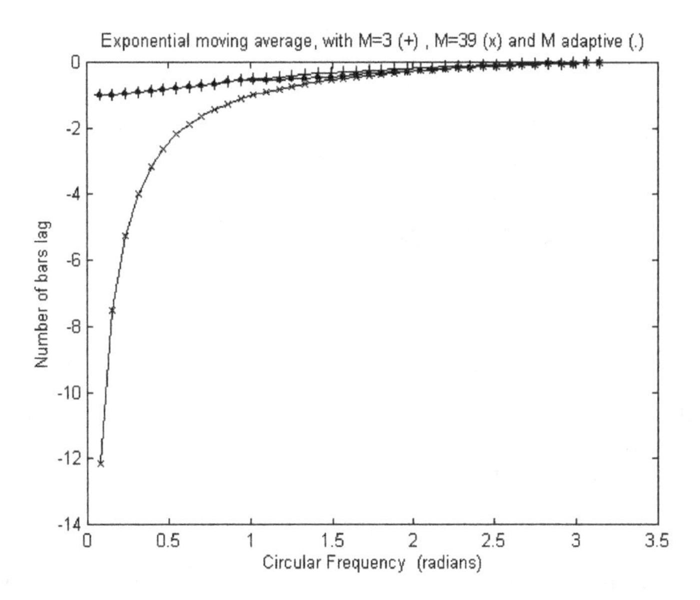

Fig 3.7(c) Number of bars lag of the exponential moving average with M = 3 (plotted as +), M = 39 (plotted as x), and an adaptive M (plotted as .) .

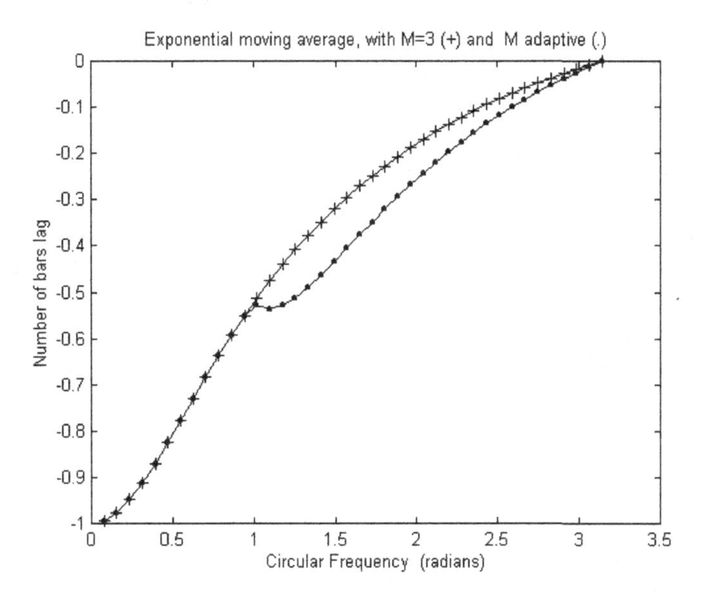

Fig 3.7(d) Number of bars lag of the exponential moving average with M = 3 (plotted as +) and an adaptive M (plotted as .).

Chapter 4

Reduced Lag Filters

We can see from the last chapter that the popular exponential moving average (EMA) has a much less phase lag than some other filters, e.g., the Butterworth filter. Thus, it would be encouraging to modify the EMA such that it can have a lesser phase lag, at least in the pass-band where signals are passed.

In 1960, Dr. R. E. Kalman applied the concept of optimal estimation to terrestrial and space navigation system. The technique has proven to be a very useful tool. Some traders have modified his filtering technique to tracking the market. Basically, they make use of a forecasted price which is a function of the current price and an estimated velocity (or slope) of the price. One modification has been created by Ehlers [2001], and is called the zero-lag EMA (ZEMA).

4.1 "Zero-lag" EMA (ZEMA)

Dr. Ehlers [2001] has simplified the Kalman filter to a single equation. The modified EMA, called the Zero-lag EMA (ZEMA) would be written as:

$$\text{ZEMA} = \alpha \times (\text{CURRENT PRICE} + \text{K V}) + (1 - \alpha) \times (\text{OLD ZEMA}) \quad (4.1)$$

where

$$\alpha = 2/(M + 1) \quad (4.2)$$

K is an adjustable parameter

V is an estimate of the velocity of market price (see, e.g., below)

(CURRENT PRICE + K V) is used to estimate what the next price would be.

More specifically, he wrote:

V = Current Price – Price of 3 bars ago

K = 0.5

$\alpha = 0.25$, which means that M = 7.

In Fig 4.1 the S & P 500 daily data are plotted together with ZEMA (thick line) and EMA with $\alpha = 0.25$ (thin line). The software used for the charting is TradeStation 2000i manufactured by Omega Research. The prices within a one-day interval is plotted as a Japanese candlestick, which looks like a candle with wicks at both ends. The body of each candle represents the absolute difference between the opening and closing prices. If the closing price is lower than the opening, the body is black. If the closing price is higher, the body is white. The tip of the upper wick represents the high within the one-day interval, and the bottom of the lower wick represents the low within the one-day interval.

The zero-lag EMA did produce less lag than the original EMA with $\alpha = 0.25$. However, to check whether ZEMA has zero lag, we need to take a look at the phase plot of its response function, which is derived in Appendix 2. The amplitude and phase of the ZEMA response function are plotted in Fig 4.2(a) and (b). From Fig 4.2(a), it can be seen that the amplitude of ZEMA is somewhat larger than that of its original EMA for almost all frequencies. From Fig 4.2(b), it can be seen that the phase lag of ZEMA is much less than that that of its original EMA for a large portion of the frequency range. However, its phase lag can be larger for a certain range of frequencies. In general, ZEMA cannot be described as zero lag, or even near zero lag.

Fig 4.1 The S & P 500 daily data are plotted together with the zero lag exponential moving average (ZEMA) (thick line) and EMA with $\alpha = 0.25$, i.e., $M = 7$ (thin line). *Chart produced with Omega Research TradeStation 2000i.*

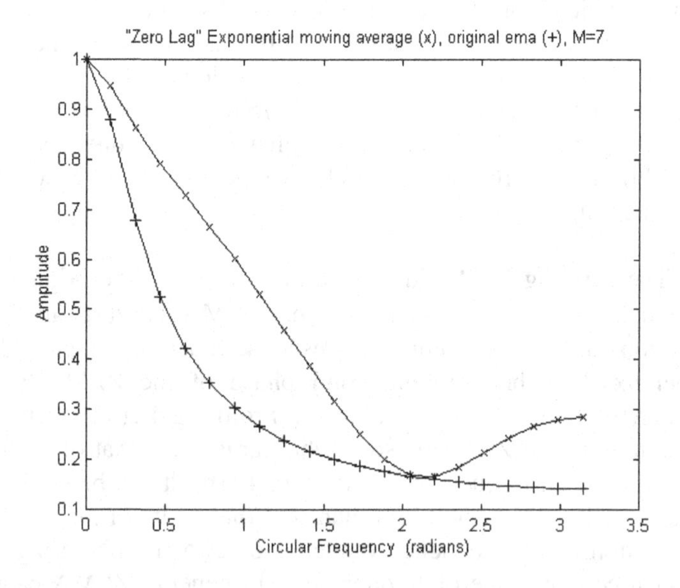

Fig 4.2(a) Amplitude response of the zero-lag exponential moving average (ZEMA) (plotted as x). Amplitude response of the exponentail moving average (EMA) with M= 7 (plotted as +) is plotted in comparison.

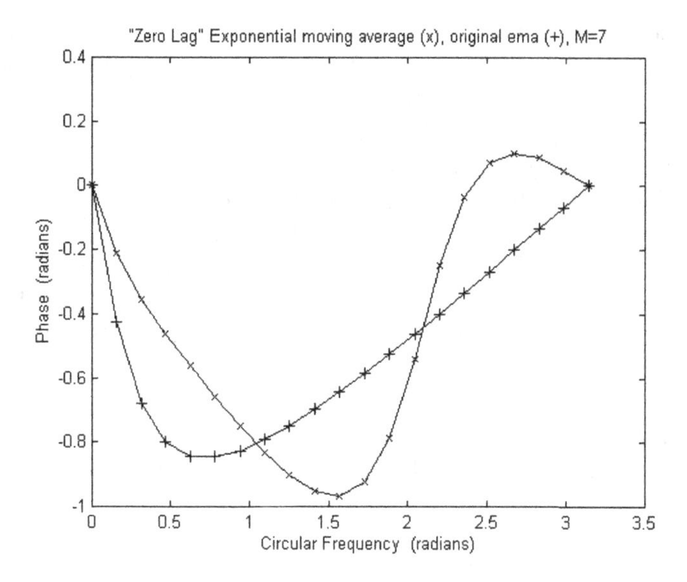

Fig 4.2(b) Phase response of the zero-lag exponential moving average (ZEMA) (plotted as x). Phase response of the exponentail moving average (EMA) with M= 7 (plotted as +) is plotted in comparison.

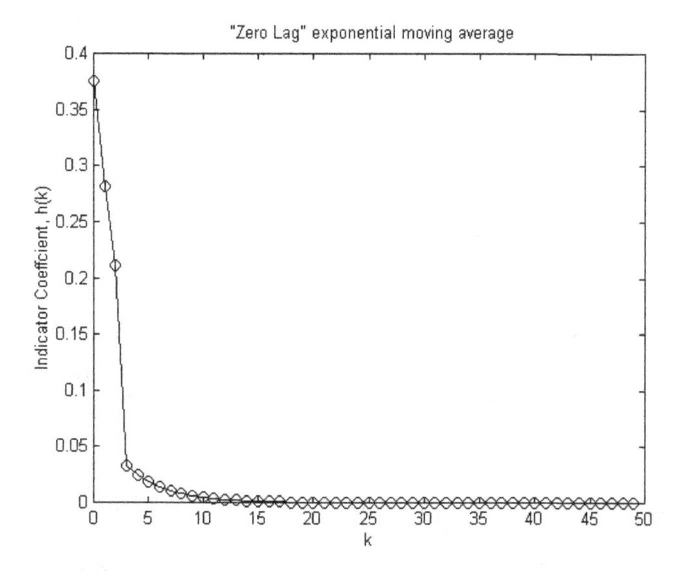

Fig 4.2(c) The indicator coefficients, h(k), of the zero-lag exponential moving average (ZEMA).

ZEMA, just like EMA, can be considered as an infinite impulse response filter, i.e., it has an infinite number of nonzero filter (or indicator) coefficients, h(k) [Mak, 2003]. However, for all practical purpose, it can be translated to a finite impulse response filter (see Appendix 2). The indicator coefficients of a finite impulse response can give us insight as to how an output signal is transformed from an input signal. The indicator coefficients, h(k), of ZEMA is plotted in Fig 4.2(c), where k = 0, 1, 2, 3,... It can be seen from the figure that all the coefficients are positive and the first three coefficients are much larger than the rest, meaning that the first three price data points play a much larger role in determining the ZEMA output.

In the next section, we will attempt to modify Eq (4.1) to see whether we can come up with a low pass filter which has much less phase lag.

4.2 Modified EMA (MEMA)

We modify Eq (4.1) and write

$$\text{MEMA} = \alpha \text{ x (CURRENT PRICE + V) + } (1 - \alpha) \text{ x (OLD MEMA)} \quad (4.3)$$

where

$$\alpha = 2/(M + 1) \quad (4.4)$$

V is an estimate of the velocity

M will be chosen to be 6. (CURRENT PRICE + V) is used to estimate what the next price would be. The cubic velocity indicator would be used to estimate V. Cubic velocity indicator has been described in Mak [2003] and will also be discussed in Chapter 8.

4.2.1 *Modified EMA (MEMA), with a Skip 1 Cubic Velocity*

The skip 1 cubic velocity mentioned in this section is actually exactly the same as the cubic velocity indicator. In the following two sections, we will be using skip 2 and skip 3 cubic velocity, as velocity is estimated using non-consecutive or skipped past price data. The concept of skipping came from a new idea, skipped convolution, introduced by Mak [2003]. Skipped convolution will be discussed in more detail in Chapter 9.

Eq (4.3) can be written as

$$y(n) = \alpha \{ x(n) + [11x(n)/6 - 3x(n-1) + 3x(n-2)/2 - x(n-3)/3] \} + (1-\alpha)\, y(n-1) \quad (4.5)$$

where

$y(n)$ is the output response

$y(n-1)$ is the output response of one bar ago

$x(n)$ is the closing price

$x(n-1)$ is the closing price of one bar ago

$x(n-2)$ is the closing price of two bars ago

$x(n-3)$ is the closing price of three bars ago.

Fig 4.3 shows the S & P500 data plotted with the modified EMA (thick line), which is compared with the original EMA with M = 6 (thin line). While the modified EMA does have a faster response than the original EMA, the response is not smooth. Before we attempt to improve on its smoothness, we will first take a look at the amplitude and phase of the response function of the modified EMA (Fig 4.4 (a) and (b)). From Fig 4.4(b), we can see that the modified EMA has a much less phase lag than that of the original EMA, and that makes it rather suitable for trading purpose. However, from Fig 4.4(a), we can see that it is not exactly a low pass filter, as it allows high frequencies to pass through, which makes its output response rather not smooth.

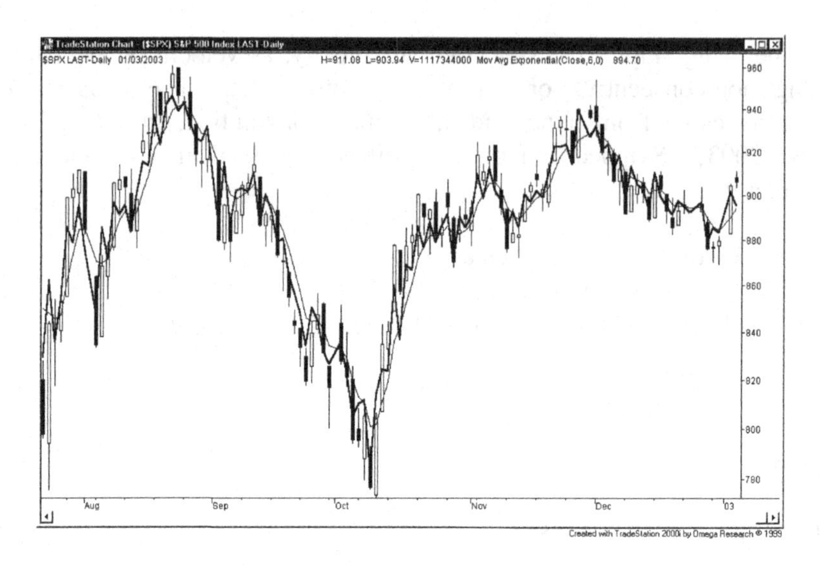

Fig 4.3 The S & P 500 daily data are plotted together with the modified exponential moving average (MEMA) with a skip 1 cubic velocity (thick line) and an EMA with M = 6 (thin line). *Chart produced with Omega Research TradeStation 2000i.*

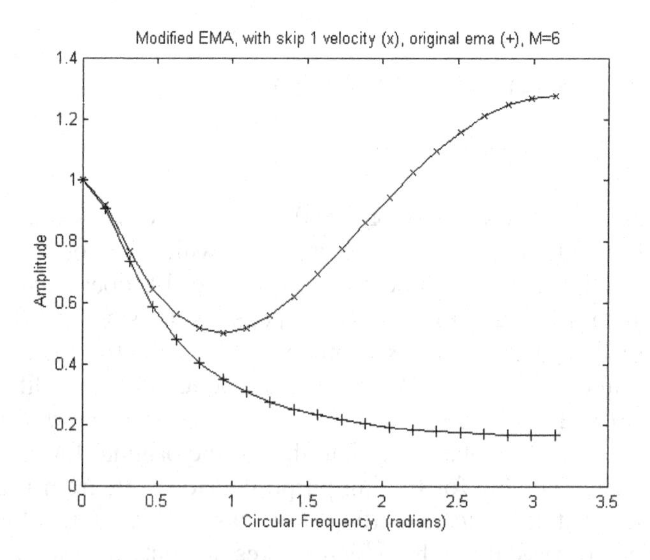

Fig 4.4(a) Amplitude response of the modified exponential moving average (MEMA) with a skip 1 cubic velocity (plotted as x). Amplitude response of the exponential moving average with M = 6 is plotted in comparison (plotted as +).

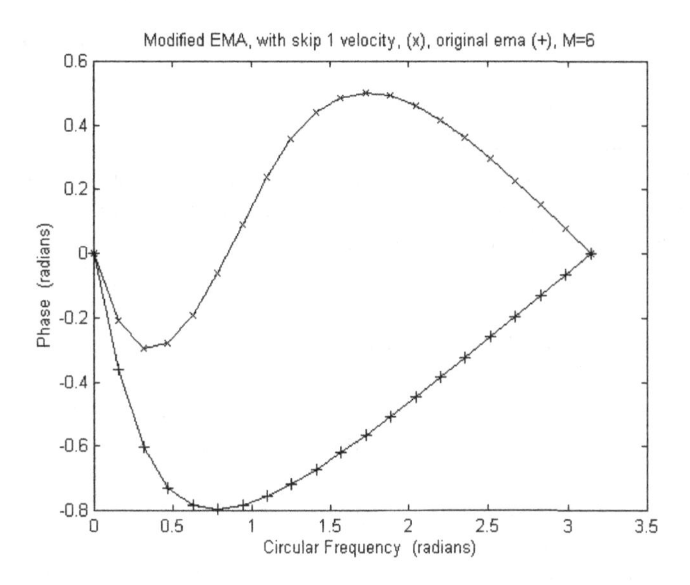

Fig 4.4(b) Phase response of the modified exponential moving average (MEMA) with a skip 1 cubic velocity (plotted as x). Phase response of the exponential moving average with M = 6 is plotted in comparison (plotted as +).

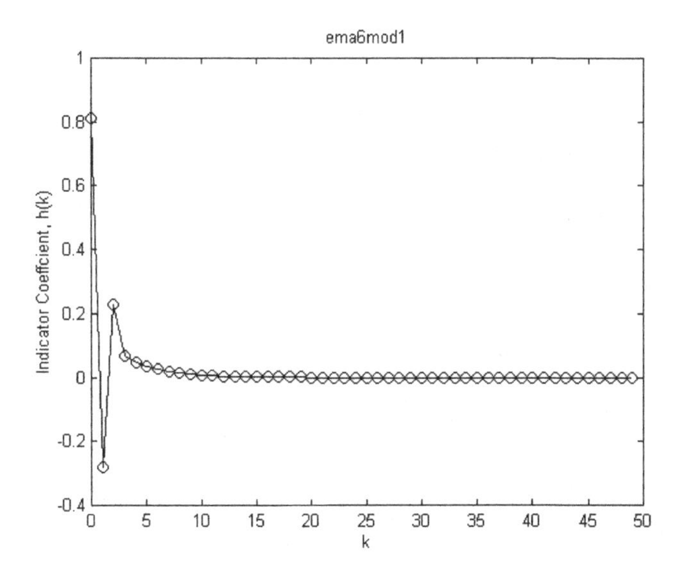

Fig 4.4(c) The indicator coefficients, h(k), of the modified exponential moving average (MEMA) with a skip 1 cubic velocity.

For the sake of scientific insight, we will take a look at the indicator coefficient if we translate Eq (4.5) into a finite impulse response filter (see Appendix 2). Fig 4.4(c) plots the indicator coefficients of the modified EMA. It can be noted that the second coefficient is negative. This can be compared with the indicator coefficients of other low pass filters, (e.g., EMA [Mak 2003, P119, 120]), where all coefficients are positive.

4.2.2 *Modified EMA (MEMA), with a Skip 2 Cubic Velocity*

The output response of the modified EMA, with a skip 1 velocity can be smoothed if we increase the skip of the velocity.

Eq (4.3) can be written as

$$y(n) = \alpha \{ x(n) + \tfrac{1}{2} [11x(n)/6 - 3x(n-2) + 3x(n-4)/2 - x(n-6)/3] \} + (1-\alpha)\, y(n-1) \qquad (4.6)$$

where

$y(n)$ is the output response

$y(n-1)$ is the output response of one bar ago

$x(n)$ is the closing price

$x(n-2)$ is the closing price of two bar ago

$x(n-4)$ is the closing price of four bars ago

$x(n-6)$ is the closing price of six bars ago.

Because we use a skip 2 velocity, the factor ½ in Eq (4.6) is necessary for yielding the correct velocity estimate.

Fig 4.5 shows the S & P500 data plotted with the modified EMA using a skip 2 velocity (thick line), which is compared with the original EMA with M = 6 (thin line). The modified EMA have a faster response than the original EMA. Furthermore, the response is smoother than that of the modified EMA with a skip 1 velocity. We will take a look at the amplitude and phase of the response function of this modified EMA (Fig 4.6 (a) and (b)). From Fig 4.6(a), we can see that the modified EMA is a low pass filter. From Fig 4.6(b), it can be seen the beginning part of the low frequencies being passed through has an approximately zero phase lag. However, the latter part of the low frequencies being passed through does have a larger phase lag than that of the original EMA. This, in general, would make this indicator a reasonable moving average for trading purpose. We will take a look at the indicator coefficient if we translate Eq (4.6) into a finite impulse response filter (see Appendix 2). Fig 4.6(c) plots the indicator coefficients of the modified EMA. It can be noted that the third and fourth coefficients are negative.

Fig 4.5 The S & P 500 daily data are plotted together with the modified exponential moving average (MEMA) with a skip 2 cubic velocity (thick line) and an EMA with M = 6 (thin line). *Chart produced with Omega Research TradeStation 2000i.*

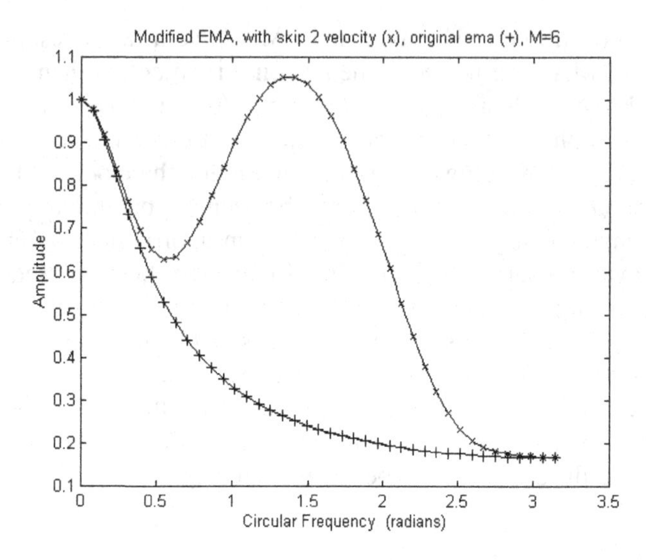

Fig 4.6(a) Amplitude response of the modified exponential moving average (MEMA) with a skip 2 cubic velocity (plotted as x). Amplitude response of the exponential moving average with M = 6 is plotted in comparison (plotted as +).

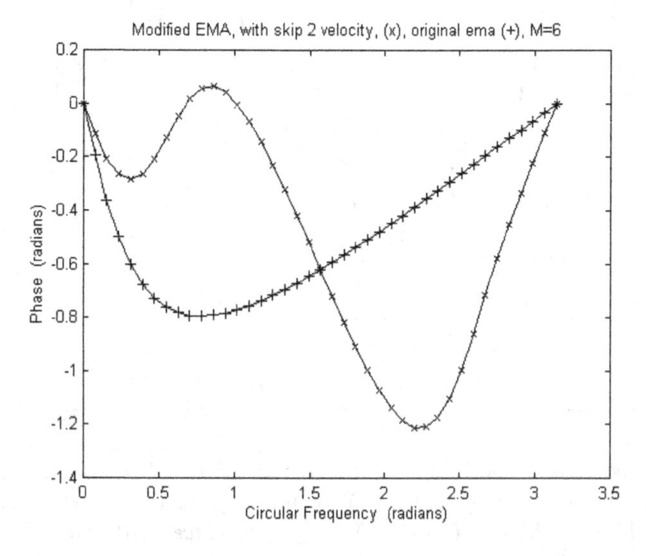

Fig 4.6(b) Phase response of the modified exponential moving average (MEMA) with a skip 2 cubic velocity (plotted as x). Phase response of the exponential moving average with M = 6 is plotted in comparison (plotted as +).

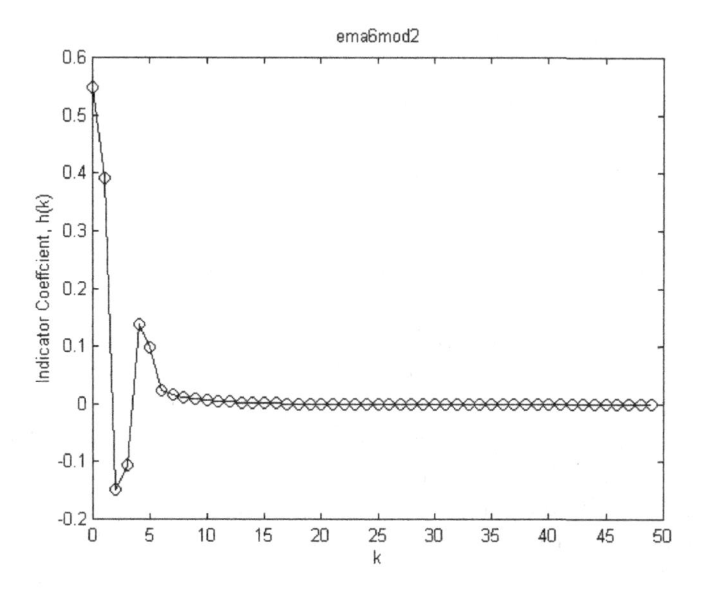

Fig 4.6(c) The indicator coefficients, h(k), of the modified exponential moving average (MEMA) with a skip 2 cubic velocity.

4.2.3 *Modified EMA (MEMA), with a Skip 3 Cubic Velocity*

The output response of the modified EMA, with a skip 2 velocity can be further smoothed if we increase the skip of the velocity even further.

Eq (4.3) can be written as

$$y(n) = \alpha \{ x(n) + 1/3 \cdot [x(n)/6 - 3x(n-3) + 3x(n-6)/2 - x(n-9)/3] \} + (1-\alpha)\, y(n-1) \qquad (4.7)$$

where

y(n) is the output response

y(n-1) is the output response of one bar ago

x(n) is the closing price

x(n-3) is the closing price of three bar ago

x(n-6) is the closing price of six bars ago

x(n-9) is the closing price of nine bars ago.

Because we use a skip 3 velocity, the factor 1/3 in Eq (4.7) is necessary for yielding the correct velocity estimate.

Fig 4.7 shows the S & P500 data plotted with the modified EMA using a skip 3 velocity (thick line), which is compared with the original EMA with M = 6 (thin line). The modified EMA have a faster response than the original EMA. Furthermore, the response is smoother than that of the modified EMA with a skip 2 velocity. We will take a look at the amplitude and phase of the response function of this modified EMA (Fig 4.8 (a) and (b)). From Fig 4.8(a), we can see that the modified EMA is a low pass filter with a lesser bandwidth than that of the modified EMA with a skip 2 velocity. From Fig 4.8(b), it can be seen the beginning part of the low frequencies being passed through has an approximately zero phase lag. However, the latter part of the low frequencies being passed through does have a larger phase lag than that of the original EMA. This, in general, would make this indicator a reasonable good moving average for trading purpose. From Fig 4.8(a), it can be seen that the amplitude has a peak when circular frequency is equal to π (i.e. 180 degrees). This can increase noise to the output. The indicator coefficient can be viewed if we translate Eq (4.7) into a finite impulse response filter (see Appendix 2). Fig 4.8(c) plots the indicator coefficients of the modified EMA. It can be noted that the fourth, fifth and sixth coefficients are negative.

The exponential moving averges modified with cubic velocity do have certain advantages. Phase lag in their low pass bands are generally less than those of the original exponential moving average. Comparing the amplitude and phase responses of the MEMA with a skip 1, 2 and 3 cubic velocity, it appears that the one with a skip 2 cubic velocity would provide the smoothest low pass filter for trading purposes.

Fig 4.7 The S & P 500 daily data are plotted together with the modified exponential moving average (MEMA) with a skp 3 cubic velocity (thick line) and an EMA with M = 6 (thin line). *Chart produced with Omega Research TradeStation 2000i.*

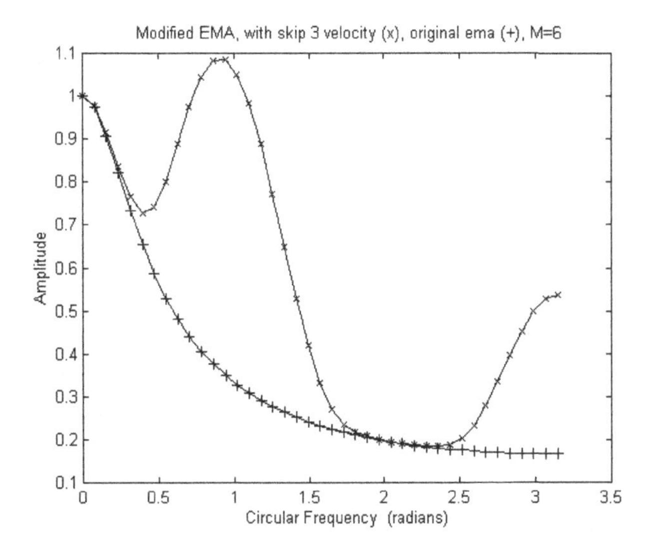

Fig 4.8(a) Amplitude response of the modified exponential moving average (MEMA) with a skip 3 cubic velocity (plotted as x). Amplitude response of the exponential moving average with M = 6 is plotted in comparison (plotted as +).

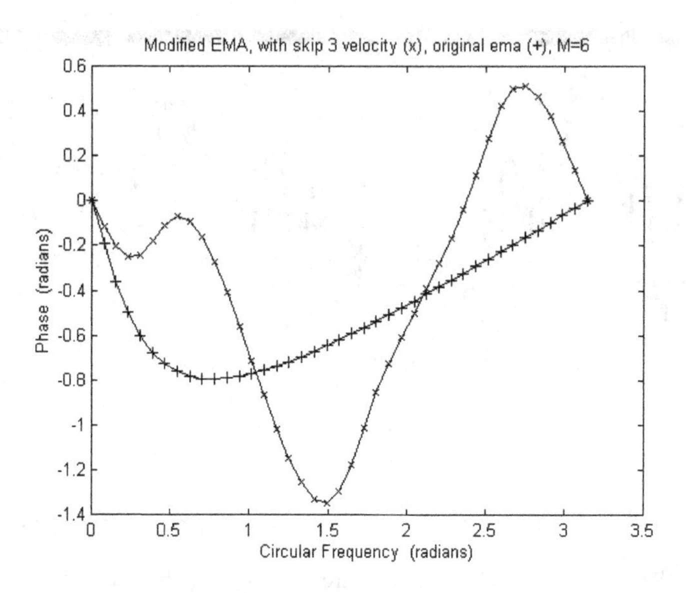

Fig 4.8(b) Phase response of the modified exponential moving average (MEMA) with a skip 3 cubic velocity (plotted as x). Phase response of the exponential moving average with M = 6 is plotted in comparison (plotted as +).

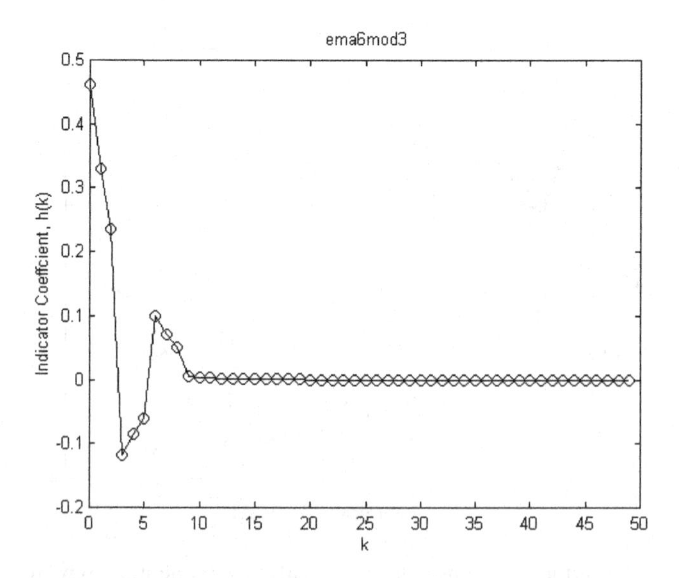

Fig 4.8(c) The indicator coefficients, h(k), of the modified exponential moving average (MEMA) with a skip 3 cubic velocity.

4.2.4 *Computer Program for Modified EMA (MEMA)*

In the EasyLanguage code of Omega Research's TradeStation2000i, the program for calculating the modified EMA can be written as follows:-

```
{*******************************************************

Description      : This Indicator plots Exponential Moving Average
that has been modified with the cubic velocity indicator which was
calculated at interval of d,  d=1,2,3,4,5,....,    d is called the skip.

*******************************************************}

Inputs: d(3),Length(6);
Plot1(XAVERAGE(c + 1/d*(11/6*c-3*c[d]+3/2*c[2*d]-1/3*c[3*d]), Length),
"plot1");
```

XAVERAGE is a build-in exponential moving average function written by TradeStation2000i. The first input parameter of XAVERAGE signifies the modified closing price series to be smoothed, while the second input parameter indicates the length, M, of the EMA. c represents the closing price of the current bar. c[d] represents the closing price of d bars ago. c[2*d] represents the closing price of 2*d bars ago, and c[3*d] represents the closing price of 3*d bars ago. d is an input parameter, and can be changed. It is taken to be 3 by default. The closing price, c, is modified by the skipped cubic velocity indicator. Length, M, of the Exponential Moving Average is taken to be 6 by default.

Chapter 5

Causal Wavelet Filters

Wavelet analysis is a mathematical tool introduced in the nineteen eighties. The technique has been discussed in detail in several books, e.g., Strang and Nguyen 1997, Rao and Bopardikar 1998, Burrus et al 1998, Daubechies 1992, Mallat 1999 and Kaiser 1994. It is particularly useful for analyzing signals of short duration. A brief summary is given in Mak [2003]. It has also been pointed out that wavelets can be an advantageous instrument for dissecting market data [Mak 2003].

Wavelets are actually bandpass filters and are adaptable to investigate market movements of long or short interval. Bandpass filters are filters that eliminate low and high frequency signals, retaining signals of middle frequencies. In that sense, they eliminate the slow trend and the noise of market actions. They can be further divided into different bands, thus giving traders the market rhythms in more details.

There are different kinds of wavelets, each one has its own father, which is called the scaling function. Scaling function is actually a low pass filter, which can yield the trend of the market. It has a cutoff frequency, ω_c . Signals above ω_c will be eliminated. However, the eliminated signals can be analyzed by wavelets. Thus, together with their father, wavelets can tell the trader what the market is up to.

The cutoff frequency, ω_c , of the scaling function can be chosen arbitrarily. To draw an analogy: imagine an onion with 12 layers. It can be divided into a center core of 4 layers and then other outer layers, or it can be divided into a center core of 6 layers, and other outer layers, or ... The center core is the scaling function, and the outer layers are the wavelets. In the terminology of electrical engineering, if there is a brick wall filter ranging in circular frequency from 0 to π, it can be divided into a low pass filter ranging in circular frequency from 0 to $\pi/8$ and then other bandpass filters, or it can be divided into a low pass filter ranging from 0 to $\pi/4$ and then other bandpass filters, or ...

Sinc wavelets are one kind of wavelets, and have been described in Mak [2003]. Their scaling function is described in Chapter 3 of the present book. In this chapter, we will describe another popular wavelets called the Mexican Hat wavelets. As will be shown later, the causal Mexican Hat wavelet has much fewer filter coefficients than the sinc wavelets, thus making it a more convenient tool for computational purpose. The Mexican Hat scaling function does not have an analytical form [Mallat 1999], and would not be investigated here.

5.1 Mexican Hat Wavelet

Mexican Hat wavelet can be expressed as

$$\psi(t) = (1-2t^2)\exp(-t^2) \tag{5.1}$$

This wavelet, different from the sinc wavelet, has the advantage that it has a compact support in time, t, i.e., it spans in time with finite duration. It is obtained by taking the second derivative of the negative Gaussian function $\exp(-t^2/2)$. $\psi(t)$ is plotted in Fig 5.1. The Fourier Transform, $\Psi(\omega)$, of $\psi(t)$ is given by [Rao 1998 P13, Mallat 1999 P80].

$$F\{\psi(t)\} = \Psi(\omega) = \int_{-\infty}^{\infty} \psi(t) e^{-j\omega t} dt = \frac{\sqrt{\pi}}{2} \omega e^{-\omega^2/4} \tag{5.2}$$

Amplitude of $\Psi(\omega)$ is plotted in Fig 5.2. It can be seen that the wavelet is a band-pass function. A function is a band-pass function if its Fourier Transform is confined to a frequency interval $\omega_1 < |\omega| < \omega_2$, where $\omega_1 > 0$ and ω_2 is finite. The Fourier Transform of the Mexican Hat wavelet peaks at exactly 2 radians. Thus the wavelet provides a filter which centers at 2 radians.

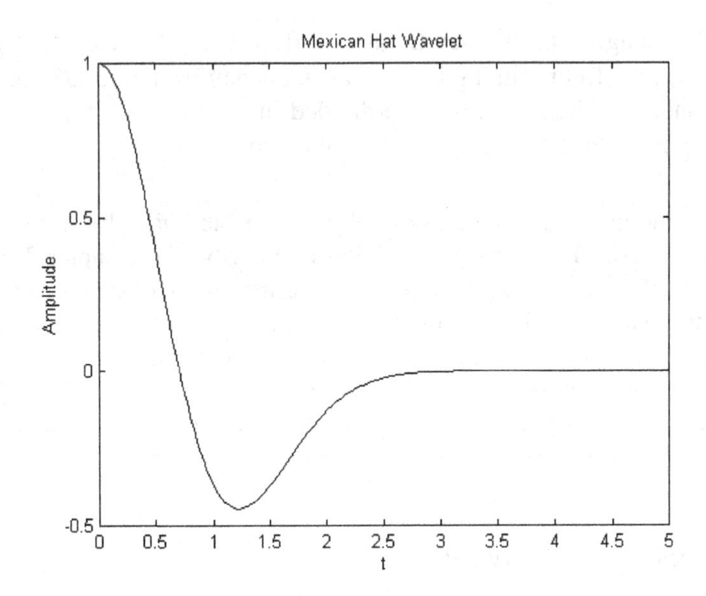

Fig 5.1 Mexican Hat wavelet in the time domain.

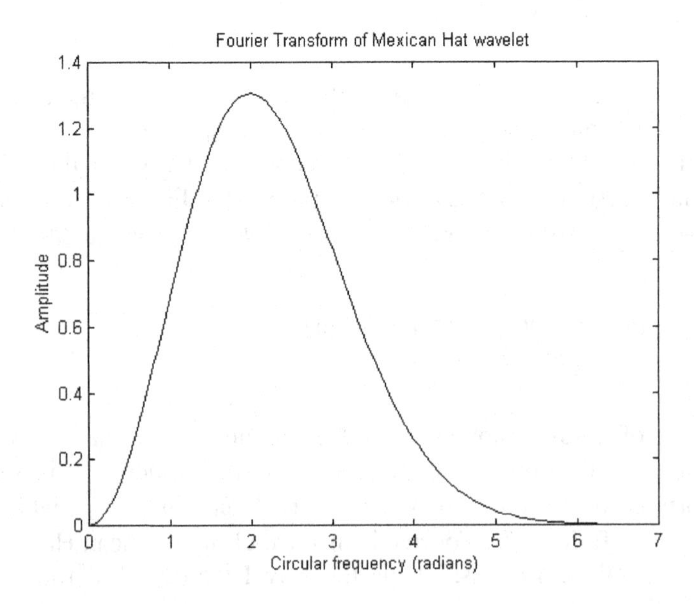

Fig 5.2 Amplitude of the Fourier Transform of the Mexican Hat wavelet plotted in circular frequency ω from 0 to 2π..

5.2 Dilated Mexican Hat Wavelet

The Fourier Transform of a wavelet which is dilated by a factor of 'a' is given by [Rao and Bopardikar 1998, Brigham 1974].

$$F\{\psi(t/a)\} = |a|\Psi(a\omega) \tag{5.3}$$

where $\psi(t)$ is called the mother wavelet,
$|a|$ is the absolute value of a.

The center frequency of $F\{\psi(t/a)\}$ is $1/|a|$ times the center frequency of the Fourier Transform of the mother wavelet, $F\{\psi(t)\}$. When $\psi(t)$ is the Mexican Hat wavelet, its Fourier Transform centers at a frequency of 2 radians. Therefore, the Fourier Transform of its dilated waveform centers at $2/|a|$ radians. Thus, the larger the value of $|a|$, the smaller the center frequency is.

The magnitude of the frequency response, i.e., the absolute value of the LHS of Eqn (5.3), for different values of a, of the Mexican Hat Wavelet, are shown in Fig 5.3. The magnitude of the peaked frequencies, when divided by the respective $|a|$, are equal to each other.

5.3 Causal Mexican Hat Wavelet

It should be noted that the RHS of Eqn (5.2) is real, i.e., there is no phase shift for all frequencies. This is because $\psi(t)$ is integrated in t from $-\infty$ to $+\infty$. In real time data analysis, since we have only past data and no future data, the data would exist only from $-\infty$ to 0. Referring to the convolution formula, this will translate into integrating $\psi(t)$ in t from 0 to $+\infty$ [Proakis and Manolakis 1996]. The Fourier Transform, $\Psi_1(\omega)$ will form a causal filter, and will be given by

$$\Psi_1(\omega) = \int_0^{\infty} \psi(t)e^{-j\omega t}dt \tag{5.4}$$

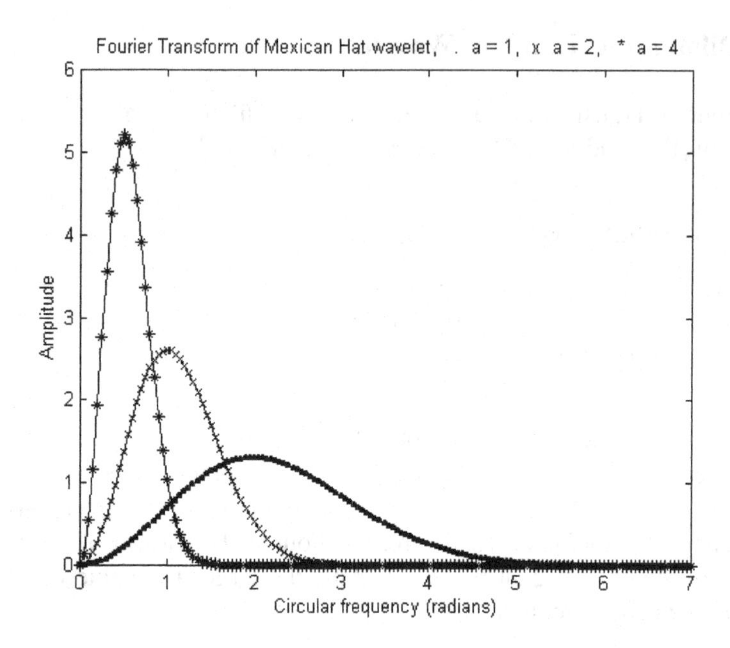

Fig 5.3 Amplitude of the Fourier Transform of the Mexican Hat wavelet for a = 1 (plotted as .), a = 2 (plotted as x) and a = 4 (plotted as *).

Substituting (5.1) into (5.4) for the Mexican Hat Wavelet, Eqn (5.4) can be expressed as

$$\Psi_1(\omega) = \frac{\sqrt{\pi}}{4}\omega^2 e^{-\omega^2/4} + j\left(\omega e^{\omega^2/4} - \int_0^\infty e^{-t^2}\sin\omega t dt\right) = R + jI \qquad (5.5)$$

where R and I are the real part and imaginary part of $\Psi_1(\omega)$ respectively.

The phase of $\Psi_1(\omega)$ is given by

$$\phi_1(\omega) = \tan^{-1}\left(\frac{I}{R}\right) \qquad (5.6)$$

This would mean that if I = 0, $\phi_1(\omega) = 0$. Referring to Eqn (5.5), this implies the phase of $\Psi_1(\omega)$ is zero when

$$\omega e^{\omega^2/4} = \int_0^\infty e^{-t^2} \sin \omega t \quad dt \tag{5.7}$$

Thus, market price signal of that particular frequency ω will experience a zero phase shift (i.e., no time delay) after being filtered by the Mexican Hat wavelet. As the center frequency of a Fourier Transform of a wavelet can be shifted by dilating the wavelet, the circular frequency ω, which corresponds to zero phase shift, can be varied.

This would mean that if we are interested in filtering out a particular frequency in a signal, we can vary 'a' in Eqn (5.3) such that the filter can output that frequency with a zero phase shift. This method has been suggested in the application of the sinc wavelet filter [Mak 2003].

5.4 Discrete Fourier Transform

We have been dealing with continuous Fourier Transform. As market data are discrete data, we need to use discrete Fourier Transform, which is given by

$$H(\omega) = \sum_0^\infty h(n) e^{-jn\omega} \tag{5.8}$$

The filter coefficients, h(n), would replace $\psi(t)$ for t = n. When $\psi(t)$ is dilated to $\psi(t/a) = \psi(t')$, the dilated filter coefficients, $h_a(n)$ would be equal to $\psi(t')$ when t' = n. For example, when a = 2, $\psi(t)$ in Fig 5.1 will be stretched horizontally by a factor of 2 to $\psi(t')$.
Choosing t' = 0, 1, 2, 3, ...

$$h_2 = (1 \;\; 0.3896 \;\; -0.3679 \;\; -0.3689 \;\; -0.1282 \;\; -0.0222 \;\; -0.0021 \;\; -0.0001 \;\; 0.0 \;\; 0.0 \; \ldots..) \tag{5.9}$$

The h_2's are shown in Fig 5.4

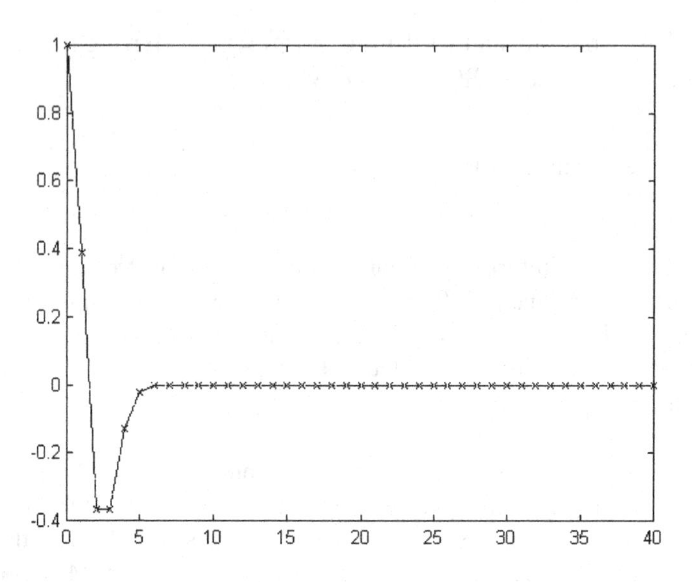

Fig 5.4 The filter coefficients $h_2(n)$ of the discrete Mexican Hat wavelet with a = 2, plotted against n.

From Eqn (5.8), the phase of H(ω) is given by

$$\phi_H(\omega) = \tan^{-1}\left(\frac{-\sum_n h(n)\sin(n\omega)}{\sum_n h(n)\cos(n\omega)} \right) \tag{5.10}$$

$\phi_H(\omega)$, for a = 2 (i.e., when $h = h_2$) is plotted in Fig 5.5 as dots. It can be observed that it begins at zero phase at ω = 0, and terminates at zero phase at ω = π. This is called a minimum-phase system [Proakis and Manolakis 1996]. Between ω= 0 and ω = π, the phase becomes zero at a frequency which we will call ω_0. The phase of the continuous Fourier Transform, $\phi_1(\omega)$, given by Eqn (5.6) is also plotted in Fig 5.5 for comparison purpose. $\phi_1(\omega)$ also has a zero phase at an ω close to ω_0. But as it is calculated from a continuous Fourier Transform, it has a different value from ω_0.

The amplitude of H(ω) in Eqn (5.8) for a = 2 for the Mexican Hat wavelet is plotted in Fig 5.6 as dots. The amplitude of the

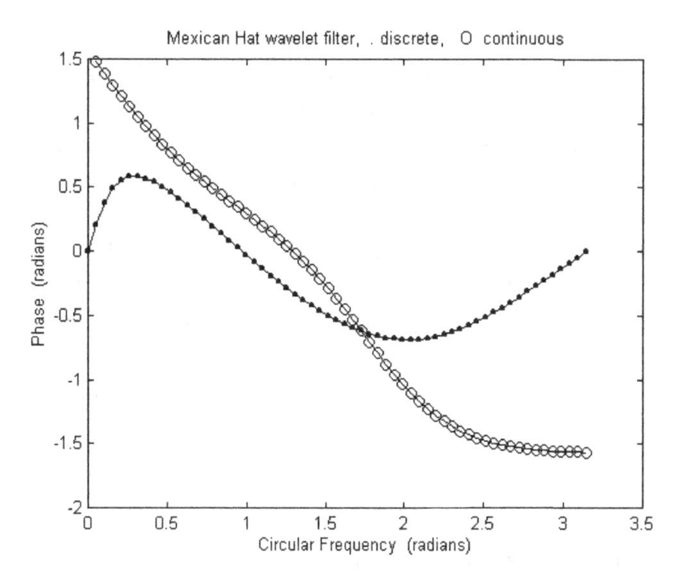

Fig 5.5 The phase of the discrete Mexican Hat wavelet (plotted as .). The phase of the continuous Mexican Hat wavelet is plotted here as comparison (plotted as o).

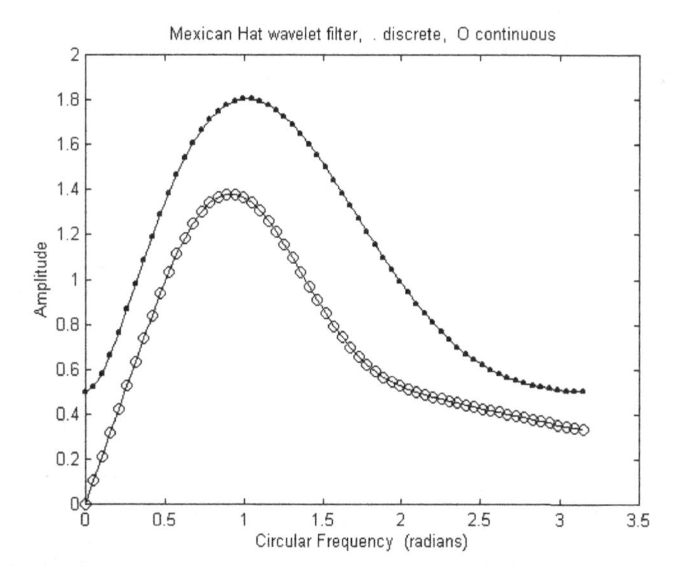

Fig 5.6 The amplitude of the discrete Mexican Hat wavelet (plotted as .). The amplitude of the continuous Mexican Hat wavelet is plotted here as comparison (plotted as o).

continuous Fourier Transform $\Psi_1(a\omega)$ is calculated from Eqn (5.5). It is then multiplied by |a| in accordance with Eqn (5.3), and plotted in Fig 5.6 for comparison purpose. It has approximately the same shape as the amplitude of $H(\omega)$. However, $H(\omega)$ has an undesirable amplitude of 0.5 for a signal whose frequency is $\omega = 0$. The factor 0.5 can be obtained by substituting the filter coefficients in Eqn (5.9) into Eqn (5.8) with the frequency $\omega = 0$. This means that the real time discrete Mexican Hat wavelet filter, while mostly functioning as a band-pass filter, cannot block out some of the low frequencies. This is a disadvantage. Nevertheless, it should be noted that the ω which corresponds to the maximum amplitude of $H(\omega)$ in Fig 5.6 approximately equals to the ω which has zero phase in Fig 5.5. This is a very advantageous feature of the filter.

5.5 Calculation of Zero Phase Frequencies

The phases for a = 1, 2, 4 and 8 calculated using Eqn (5.10) are plotted in Fig 5.7. The frequencies, ω_0's, which correspond to zero phases for different a's, can be exactly located by using numerical analysis. They are listed in Table 5.1. The ω_1's, which correspond to the maximum amplitude of $H(\omega)$'s for different a's can also be found using numerical analysis. They are also listed in Table 5.1.

Table 5.1

a	2/a	ω_0 (radians) (zero phase)	ω_1 (radians) (maximum amplitude)
64	0.03125	0.0289	0.0345
32	0.0625	0.0578	0.0689
16	0.125	0.1156	0.1371
8	0.25	0.2316	0.2712
4	0.5	0.4670	0.5311
2	1	0.9686	1.0198
3/2	4/3	1.3558	1.3233

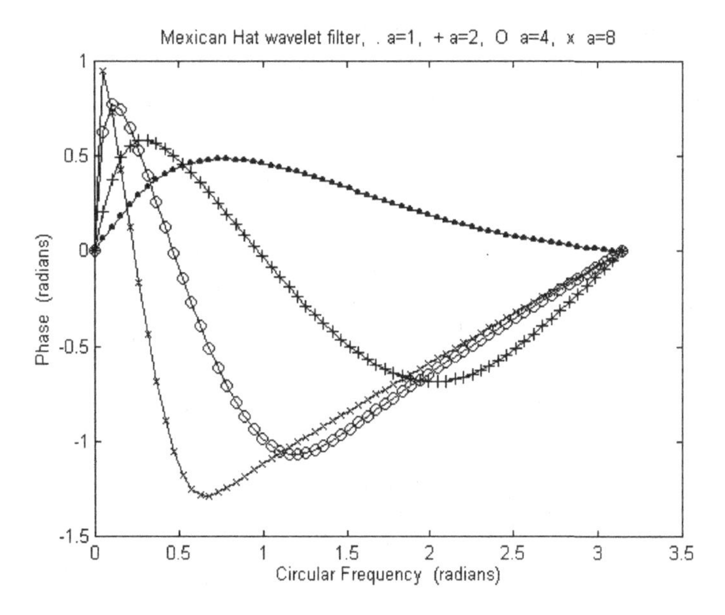

Fig 5.7 The phase of the discrete Mexican Hat wavelet for a = 1 (plotted as .), a = 2 (plotted as +), a = 4 (plotted as o) and a = 8 (plotted as x).

Had the filter been non-causal and continuous as in Eqn (5.2), the zero phase ω_0's would be exactly equal to 2/a, and the Fourier Transform would attain maximum amplitude at that frequency. However, because of its causality, it can be seen from Table 5.1 that the zero phase ω_0 is only approximately equal to 2/a. Furthermore, the zero phase ω_0's are approximately equal to the ω_1's where the filter attains maximum amplitudes. In spite of these, the Mexican Hat wavelet can make a very good band-pass filter. As the zero phase ω_0 is approximately equal to 2/a, this would imply that if we know the frequency of a signal, then we can find 'a' of the causal discrete Mexican hat wavelet filter which will render a zero phase shift after the signal is filtered. In order to determine 'a' more accurately, 2/a is curve fitted to the frequency ω_0. The fitted 2/a, $(2/a)_f$, is given by Eqn (5.11)

$$(2/a)_f = 1.091\omega_0 - 0.071\omega_0^2 \tag{5.11}$$

It is is listed in column 3 of Table 5.2. a_f can then be calculated as 2 times the reciprocal of column 3 and is listed in column 4. It can be seen that a_f is approximately equal to a, which is listed in column 5.

Table 5.2

ω_0(radians) (zero phase)	2/a	$(2/a)_f = 1.091\omega_0 - 0.071\omega_0^2$	a_f	a
0.0289	0.03125	0.031478	63.54	64
0.0578	0.0625	0.062837	31.83	32
0.1156	0.125	0.1252	15.97	16
0.2316	0.25	0.2489	8.035	8
0.4670	0.5	0.494	4.048	4
0.9687	1	0.990	2.02	2
1.3558	1.3333	1.349	1.483	1.5

As the amplitude of the filtered signal is different from that of the original signal, we would like to normalize the discrete Fourier Transform such that the filtered signal would have the same amplitude as the original signal for frequency ω_0. The amplitude, $|H(\omega_0)|$, which corresponds to the ω_0 with zero phase shift, is listed in column 3 of Table 5.3. This amplitude is calculated when we arbitrarily assumed that the number of filter coefficients could be truncated at 41. Increasing the number of filter coefficients would only have slightly changed the amplitude. $|H(\omega_0)|$ are curve fitted to a_f, yielding an $|H(\omega_0)|_f$:

$$|H(\omega_0)|_f = 0.488 + 0.646\, a_f + 0.0001\, a_f^2 \qquad (5.12)$$

This is listed in column 4 of Table 5.3. They compared reasonably with $|H(\omega_0)|$ listed in column 3. An input signal of unit amplitude of frequency ω_0 will yield an output signal of amplitude $|H(\omega_0)|$ after filtering. To normalize any output signal of frequency ω_0, the amplitude of the filtered signal will be divided by $|H(\omega_0)|_f$, where $|H(\omega_0)|_f$ is calculated from Eqn (5.12).

Table 5.3

ω_0 (radians) (zero phase)	a_f	$\lvert H(\omega_0)\rvert$	$\lvert H(\omega_0)\rvert_f$
0.0289	63.54	41.75	41.95
0.0578	31.83	21.12	21.16
0.1156	15.97	10.81	10.83
0.2316	8.035	5.66	5.69
0.4670	4.048	3.08	3.11
0.9687	2.02	1.80	1.79
1.3558	1.483	1.48	1.45

5.6 Examples of Filtered Signals

We will now look at the output response of a real time Mexican Hat wavelet filter when the input signal has one or more than one frequencies. We would assume that we know the frequencies of the input signal and would like the signal of one of the frequencies, $\omega_{0,}$, to be filtered out with a zero phase shift (i.e., no time lag). Knowing ω_0, the computer program will calculate a_f from Eqn (5.11) and $\lvert H(\omega_0)\rvert_f$ from Eqn (5.12). The filter coefficients, $h_a(n) = \psi(t/a_f = n)$ are calculated and used to convolute with the input signal. The number of coefficients needed can be set to 4 x round(a_f) as the other coefficients are approximately zero in value and would hardly affect the computation (see Fig. 5.4). The functional value of round(a_f) is equal to an integer after a_f is being rounded off.

5.6.1 *Signal with Frequency* $\pi/4$

An input signal of price, $p = \sin(n\pi/4)$, is plotted as 'o' in Fig 5.8. It was then filtered by the Mexican Hat wavelet. The filtered signal is plotted as 'x' in the same figure. The two signals almost overlap each other. This implies that the filtered output signal has negligible phase shift from the input signal.

As described in Mak [2003], the first and second derivatives can be used to forecast the direction of the market. The cubic velocity and acceleration indicators can simulate the first and second derivative respectively [Mak 2003]. We will therefore take a look at the two derivatives. The first derivative of p, i.e. $(\pi/4)\cos(n\pi/4)$, is plotted in Fig 5.9(a) and compared with the velocity obtained by operating on the filtered signal with the cubic velocity indicator. (Cubic velocity indicator is described in Chapter 8). The agreement is reasonably well. The second derivative of p, i.e., $-(\pi/4)^2\sin(n\pi/4)$, is plotted in Fig 5.9(b) and compared with the acceleration obtained by operating on the filtered signal with the cubic acceleration indicator. (Cubic acceleration indicator is described in Chapter 8). The agreement is still reasonable.

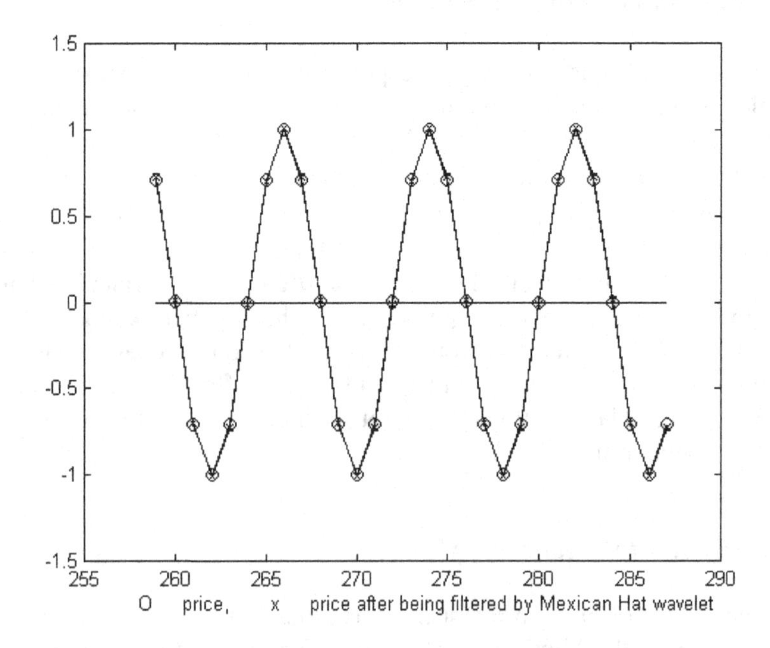

Fig 5.8 An input signal of price, $p = \sin(n\pi/4)$, is plotted as 'o'. The signal is filtered by the Mexican Hat wavelet and the output response is plotted as 'x'.

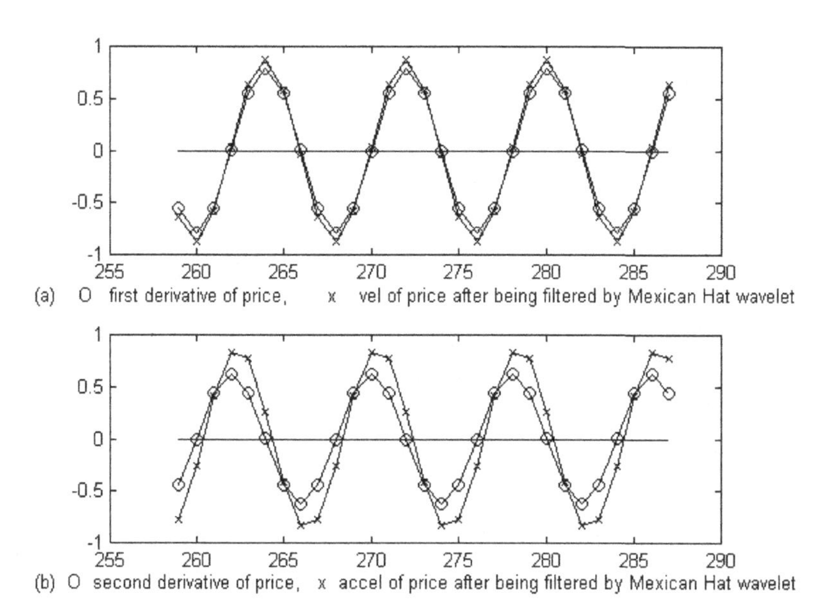

Fig 5.9(a) The first derivative of p, i.e. $(\pi/4)\cos(n\pi/4)$, is plotted as 'o' and compared with the velocity (plotted as x) obtained by operating on the Mexican Hat wavelet filtered signal with the cubic velocity indicator.
(b) The second derivative of p, i.e., $-(\pi/4)^2\sin(n\pi/4)$, is plotted as 'o' and compared with the acceleration (plotted as x) obtained by operating on the Mexican Hat wavelet filtered signal with the cubic acceleration indicator.

5.6.2 *Signal with Frequency $\pi/32$*

We will try the filter on an input signal with a lower frequency. An input signal of price, $p = \sin(n\pi/32)$, is plotted as 'o' in Fig 5.10. It was then filtered by the Mexican Hat wavelet. The filtered signal is plotted as 'x' in the same figure. The two signals almost overlap each other. This implies that the filtered output signal has negligible phase shift from the input signal.

The first derivative of p, i.e. $(\pi/32)\cos(n\pi/32)$, is plotted in Fig 5.11(a) and compared with the velocity obtained by operating on the filtered signal with the cubic velocity indicator. The two signals almost overlap each other. The second derivative of p, i.e., $-(\pi/32)^2\sin(n\pi/32)$,

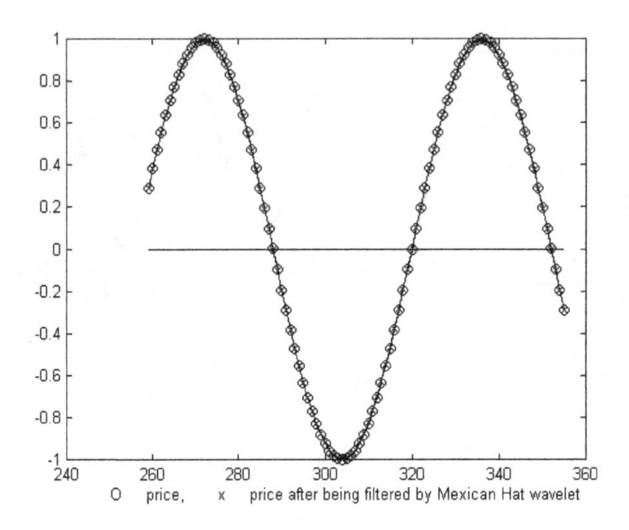

Fig 5.10 An input signal of price, p = sin(nπ/32), is plotted as 'o' . The signal is filtered by the Mexican Hat wavelet and the output response is plotted as 'x'.

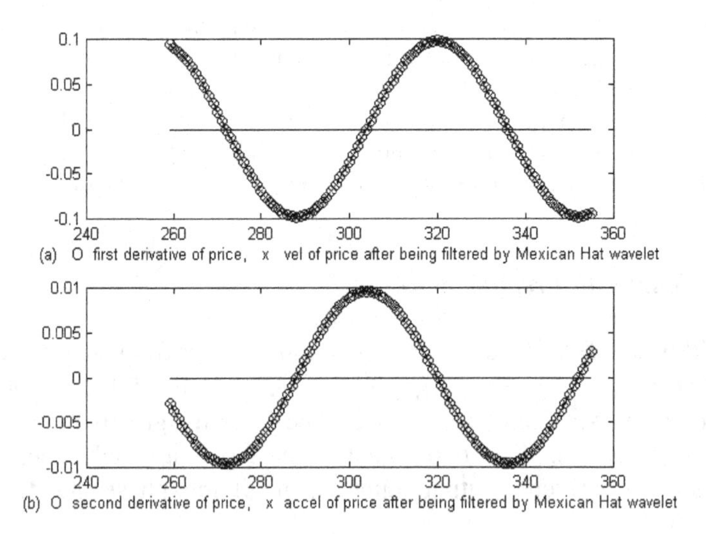

Fig 5.11(a) The first derivative of p, i.e. $(\pi/32)\cos(n\pi/32)$, is plotted as 'o' and compared with the velocity (plotted as x) obtained by operating on the Mexican Hat wavelet filtered signal with the cubic velocity indicator.

(b) The second derivative of p, i.e., $-(\pi/32)^2\sin(n\pi/32)$, is plotted as 'o' and compared with the acceleration (plotted as x) obtained by operating on the Mexican Hat wavelet filtered signal with the cubic acceleration indicator.

is plotted in Fig 5.11(b) and compared with the acceleration obtained by operating on the filtered signal with the cubic acceleration indicator. Again, the two signals almost overlap each other. The agreement is much better than the higher frequency of $\pi/4$ because the cubic velocity and cubic acceleration indicators have much less phase shifts at lower frequencies.

5.6.3 *Signal with Frequencies $\pi/4$, and $\pi/32$*

We will try the filter on an input price, p, with a high frequency signal p_1 superimposed on a low frequency signal p_2 with larger amplitude.

$$p = p_1 + p_2 = \sin(n\pi/4) + 5\sin(n\pi/32) \tag{5.13}$$

The low frequency component will be eliminated by the Mexican Hat wavelet. Fig 5.12(a) plots p_1 , p_2 and the summation of p_1

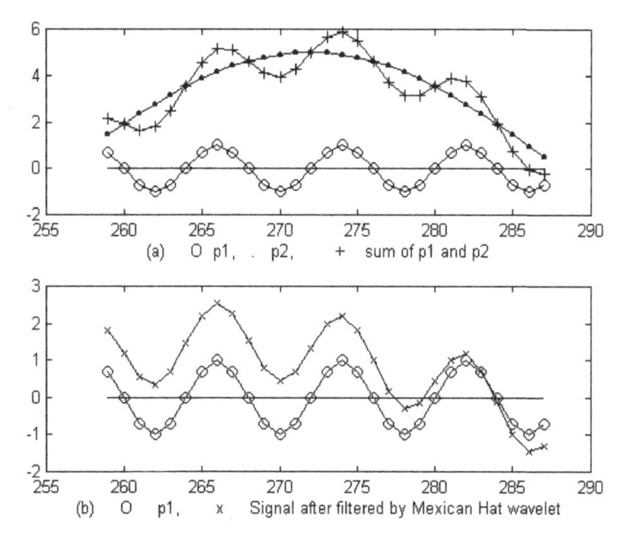

Fig 5.12(a) An input signal of price, $p = p_1 + p_2 = \sin(n\pi/4) + 5\sin(n\pi/32)$, is plotted as '+' . The high frequency component, p_1 , is plotted as 'o' and the low frequency component , p_2 , is plotted as '.' .
(b) The input signal is filtered by the Mexican Hat wavelet designed to eliminate the low frequency component. The output response is plotted as 'x'. The low frequency component of the original input signal, p_1 , is plotted as 'o' for comparison.

and p_2. Fig 5.12(b) shows the signal after being filtered. The frequency component p_1 is plotted as comparison. Some of the low frequency component has not been filtered off. This is comprehensible as the real time discrete Mexican Hat filter cannot filter off some of the low frequencies (see Fig 5.6).

The first derivative of p_1, i.e. $(\pi/4)\cos(n\pi/4)$, is plotted in Fig 5.13(a) and compared with the velocity obtained by operating on the filtered signal with the cubic velocity indicator. The agreement is reasonably well. The second derivative of p_1, i.e., $-(\pi/4)^2\sin(n\pi/4)$, is plotted in Fig 5.13(b) and compared with the acceleration obtained by operating on the filtered signal with the cubic acceleration indicator. The agreement is still reasonable. The cubic velocity and acceleration indicators are actually high pass filters. They therefore are able to eliminate the low frequency component that is not filtered off by the Mexican Hat wavelet.

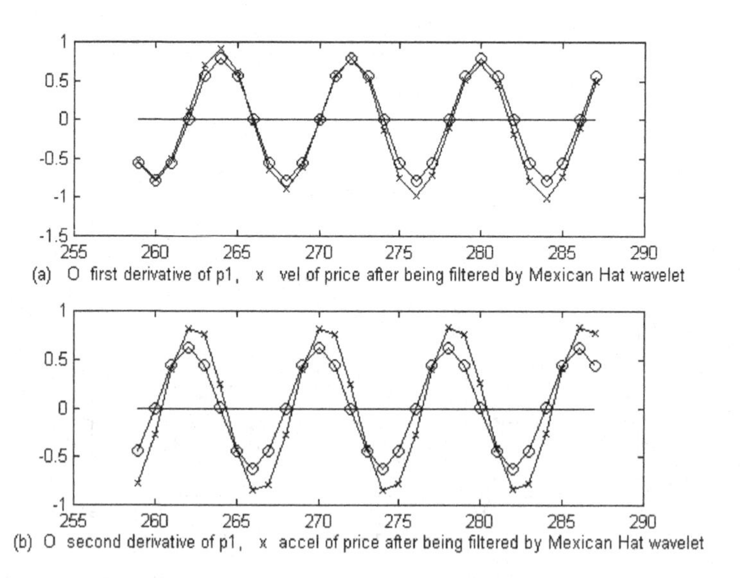

Fig 5.13(a) The first derivative of p_1 , i.e. $(\pi/4)\cos(n\pi/4)$, is plotted as 'o' and compared with the velocity (plotted as x) obtained by operating on the Mexican Hat wavelet filtered signal with the cubic velocity indicator.
(b) The second derivative of p_1 , i.e., $-(\pi/4)^2\sin(n\pi/4)$, is plotted as 'o' and compared with the acceleration (plotted as x) obtained by operating on the Mexican Hat wavelet filtered signal with the cubic acceleration indicator.

5.7 High, Middle and Low Mexican Hat Wavelet Filters

If the frequencies of the signal are not known, we can attempt to use a series of Mexican Hat Wavelet filters to filter the signal. One series can be chosen to have a = 1.483 (high), 4.048 (middle), 15.97 (low) (see Table 5.3). The filter coefficients are then normalized by dividing by $|H(\omega_0)|_f$ (see Table 5.3) and are listed as follows:

$h_{high}(n)$ = (0.6897 0.0397 -0.2951 -0.0828 -0.0065 -0.0002
0.0000)

$h_{middle}(n)$ = (0.3215 0.2656 0.1289 -0.0183 -0.1154 -0.1434
-0.1213 -0.0805 -0.0441 -0.0204 -0.0081 -0.0027
-0.0008 -0.0002 0.0000)

$h_{low}(n)$ = (0.0923 0.0913 0.0880 0.0828 0.0758 0.0673
0.0575 0.0469 0.0358 0.0245 0.0135 0.0029
-0.0068 -0.0155 -0.0230 -0.0292 -0.0341 -0.0377
-0.0399 -0.0411 -0.0411 -0.0403 -0.0387 -0.0365
-0.0339 -0.0311 -0.0280 -0.0250 -0.0220 -0.0191
-0.0164 -0.0139 -0.0117 -0.0097 -0.0080 -0.0065
-0.0053 -0.0042 -0.0033 -0.0026 -0.0020)

These coefficients are plotted in Fig 5.14. Amplitude of the Fourier Transform of these filter coefficients are plotted in Fig 5.15. Note that the amplitudes of the peaks are 1, as the filter coefficients are normalized.

5.8 Limitations of Mexican Hat Wavelet Filters

While the discrete causal Mexican Hat wavelet filters form a series of bandpass filters, they do not have very good resolutions. This, of course, has always been a problem with designing filters. As we can see from Fig 5.15, the amplitudes of their Fourier Transforms overlap each other. This would cause a signal of a single frequency to appear as a signal in all the filtered signals.

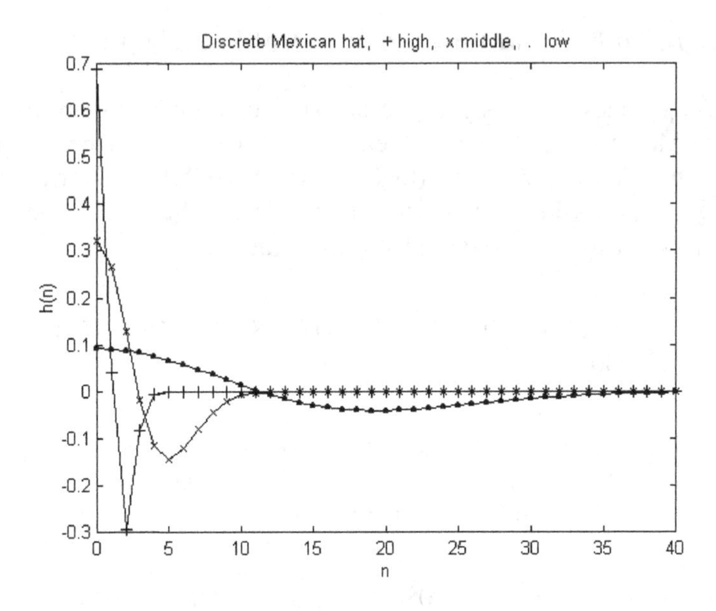

Fig 5.14 High (+), Middle (x) and Low (.) Mexican Hat Wavelet coefficients.

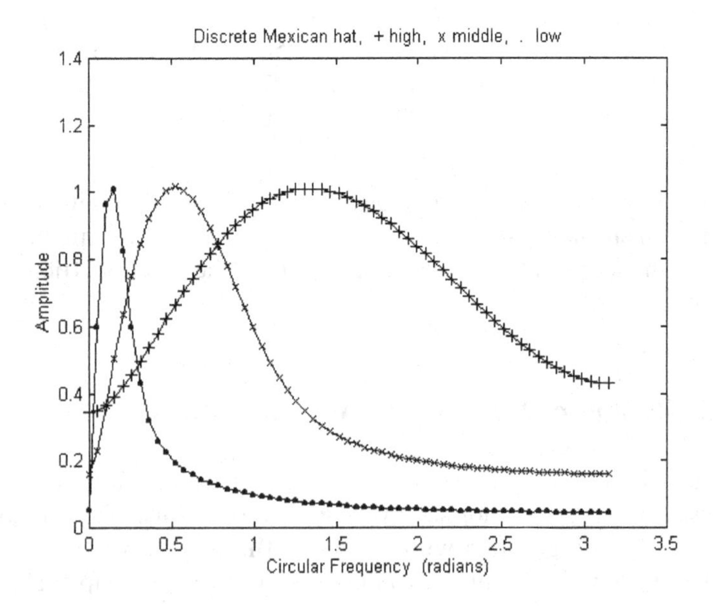

Fig 5.15 Amplitudes of the Fourier Transform of the High (+), Middle (x) and Low (.) Mexican Hat Wavelet coefficients.

Fig 5.16 shows an input sine wave with a single low circular frequency of 0.116 radians. It was filtered by the series of Wavelet filters. The output signal from the low wavelet filter (plotted as .) almost coincides with the input signal (plotted as o). However, the output signals from the high and middle wavelet filters also produce signals of lower amplitude. These signals are simply caused by their frequency bands spreading into the low frequency range, as can be observed in Fig 5.15.

Fig 5.17 shows an input sine wave with a single circular frequency of 0.467 radians. It was filtered by the series of Wavelet filters. The output signal from the middle wavelet filter (plotted as x) almost coincides with the input signal (plotted as o). However, the output signals from the high and low wavelet filters also produce signals of lower amplitude. Again, this is consistent with what is being shown in Fig 5.15.

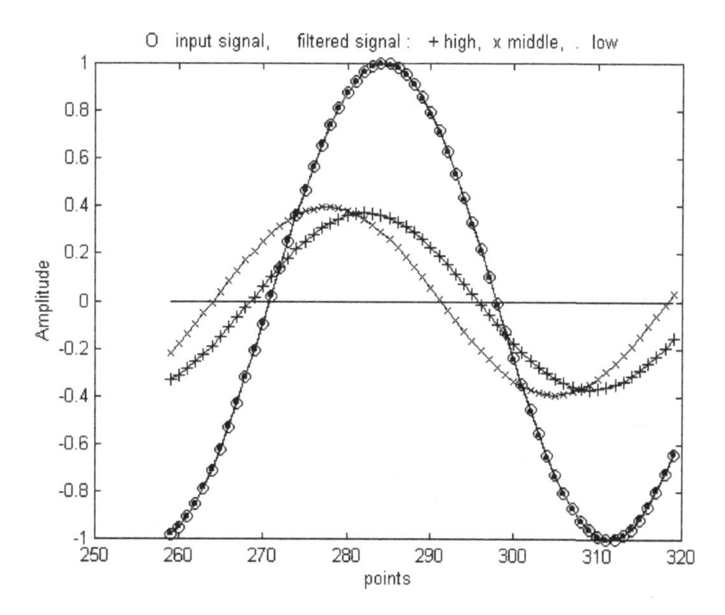

Fig 5.16 The input signal (plotted as o) has a circular frequency of 0.116 radians. It was filtered by the High, Middle and Low Wavelet filters. The output signal from the Low wavelet filter (plotted as .) almost coincides with the input signal (plotted as o). However, the output signals from the High and Middle wavelet filters also produce signals of lower amplitude.

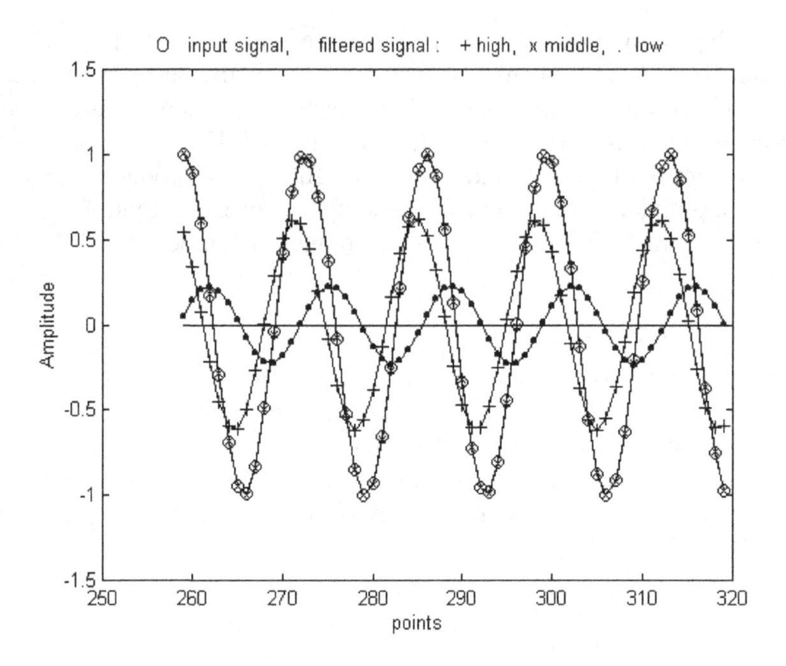

Fig 5.17 The input signal (plotted as o) has a circular frequency of 0.467 radians. It was filtered by the High, Middle and Low Wavelet filters. The output signal from the Middle wavelet filter (plotted as x) almost coincides with the input signal (plotted as o). However, the output signals from the High and Low wavelet filters also produce signals of lower amplitude.

Fig 5.18 shows an input sine wave with a single circular frequency of 1.36 radians. It does not look like a pure sine wave simply because the sampled points are not dense enough to show a pure sine wave. The signal was filtered by the series of Wavelet filters. The output signal from the high wavelet filter (plotted as +) almost coincides with the input signal (plotted as o). However, the output signals from the middle and low wavelet filters also produce signals of lower amplitude. Again, this is consistent with what is being shown in Fig 5.15.

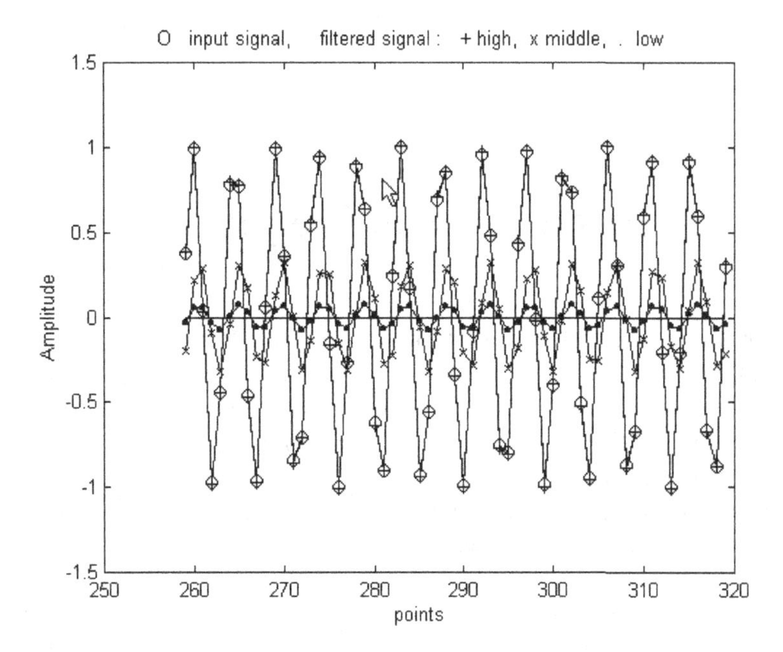

Fig 5.18 The input signal (plotted as o) has a circular frequency of 1.36 radians. It was filtered by the High, Middle and Low Wavelet filters. The output signal from the high wavelet filter (plotted as +) almost coincides with the input signal (plotted as o). However, the output signals from the Middle and Low wavelet filters also produce signals of lower amplitude.

Compared to the sinc wavelet filters [Mak 2003], the Mexican Hat wavelet filters use a much smaller number of coefficients, h(n), to convolute with the input signal. However, the sinc wavelet filters have much sharper frequency bandwidths than the Mexican Hat wavelet filters. Other wavelet filters can be used. Their amplitude and phase characteristics should be inspected to exploit their properties.

Chapter 6

Instantaneous Frequency

Traders quite often would like to know how fast the market is moving, and adjust their exponential moving average to adapt to the moving trend. One way to know how fast the market is changing is to estimate the frequency, ω, or cycle period, T ($= 2\pi/\omega$), of the market price. Knowing ω, one can vary the exponential moving average for smoothing the market movement. An example of this type of adaptive exponential moving average has been shown in Chapter 3.

In addition, some traders believe that the market comes in cycles. Thus, they would like to ride the wave up when the market ascends by going long, and ride the wave down when the market descends by selling short. Thus, there is immense interest to find out what cycle periods are in the market data [Berstein 1991, Pring 1991, Ehlers 1992, 2001].

Furthermore, as discussed in the last chapter, knowing ω, one can pick a Mexican Hat wavelet filter, or maybe other wavelet filters, or other bandpass filters (see Chapter 10), such that the output signal has a zero phase or time lag. Thus, finding ω would definitely be of great advantage to traders.

To find cycles or frequencies in market data, the mathematical tool, Fourier Analysis can be used. Unfortunately, Fourier Analysis work well only with long, regular signals,, and not well with signals of short duration [Mak 2003]. Thus, other methods have been suggested. One approach is to take an approximation, and assume that there is only one dominant cycle in the data A cycle would have a certain rate of phase change. Hilbert Transform is then used to generate the InPhase and Quadrature components, from which the phase at each bar is measured. (A bar is one data point on the chart. For example, one bar in the daily chart represents one day). The rate of phase change is determined as the differential phase from bar to bar. Three different methods have been developed, and one of the methods, the Homodyne Discriminator is considered to be the most accurate [Ehlers 2001]. However, the Homodyne Discrimator produces a cycle period measurement with a lag of 20.5 bars. For a daily chart, the trader would only know the cycle period of 20.5 days back. As market can move rather quickly, this long lag can cause the information to be outdated. We will attempt to model the market data rather differently, so that the cycle period or frequency measurement has only a lag of 3 or 3.5 bars.

6.1 Calculation of Frequency (4 data points)

We will model market price data as a single sine wave superimposed on a constant level. The equation can be written as

$$x = A \sin(\omega t + \phi) + D \tag{6.1}$$

where x is the market price,
A is the amplitude of the sine wave,
ω is the circular frequency of the sine wave,
ϕ is the phase when time $t = 0$,
D is a constant.

Eqn (6.1) has four unknowns, A, ω, ϕ and D. Their solutions would require at least four data points (x, t). Details of the solution are given in Appendix 3. Specifically, the circular frequency, ω, is given by

$$\omega = 2 \sin^{-1}\left[\frac{1}{2}\left(3 - \frac{x_0 - x_{-3}}{x_{-1} - x_{-2}}\right)^{1/2}\right] \tag{6.2}$$

where x_0 is the closing price of the current bar, i.e., at $t = 0$
x_{-1} is the closing price of one bar ago, i.e., at $t = -1$
x_{-2} is the closing price of two bars ago, i.e., at $t = -2$
x_{-3} is the closing price of three bars ago, i.e., at $t = -3$

The period, T, of the sine wave can be given as $2\pi/\omega$. Eqn (6.2) implies that ω or T can be calculated instantly, and the time delay is only 2 bars, which is half the number of data points.

6.2 Wave Velocity

The wave velocity, v, which is defined as the slope or derivative of the price, x, is given by differentiating Eqn (6.1)

$$v = A\,\omega \cos(\omega t + \phi) \tag{6.3}$$

At t = 0, the current wave velocity, v_0, can be written as

$$v_0 = A\,\omega \cos(\phi) \tag{6.4}$$

6.3 Wave Acceleration

The wave acceleration, a, which is defined as the slope of the slope or second derivative of the price, x, is given by differentiating Eqn (6.3)

$$a = -A\,\omega^2 \sin(\omega t + \phi) \tag{6.5}$$

At t = 0, the current wave acceleration, a_0, can be written as

$$a_0 = -A\,\omega^2 \sin(\phi) \tag{6.6}$$

6.4 Examples using 4 Data Points

Fig. 6.1(a) shows price data (marked as +) simulated as a sine wave plotted versus t, with $A = 0.25$, $\omega = \pi/4 = 0.7854$, $\phi = \pi/3$ and $D = 0.6$. For each instant of t, t is chosen to be 0 for the current data point in the calculation using the model of Eqn (6.1). This current data point, together with three previous data points are employed to solve for A, ω, ϕ and D. The estimated parameters are then used to calculate a current price value (marked as o in Fig 6.1(a)) using Eqn (6.1) with t = 0. The first three points are not calculated, as it required four points to calculate the frequency. It can be seen that the calculated values agree very well with the original price data. The circular frequency, ω, calculated at each point is shown in Fig 6.1(b), where a line representing the input frequency is also plotted. The calculated frequencies agree with the original frequency exactly. Fig 6.2(a) plots the velocity (marked as +) as derived from the original price data using Eqn (6.3), as well as the wave velocity (marked as o) calculated from Eqn (6.4) using the solved parameters. Fig 6.2(b) plots the acceleration (marked as +) as derived from the original price data using Eqn (6.5), as well as the wave acceleration (marked as o) calculated from Eqn (6.6) using the solved parameters. Both the calculated wave velocity and acceleration values agree with the theoretically derived data quite well. It should be noted that the calculated values has no phase shift or time lag. This makes modeling the price data with a sine wave more accurate than modeling with a polynomial function.

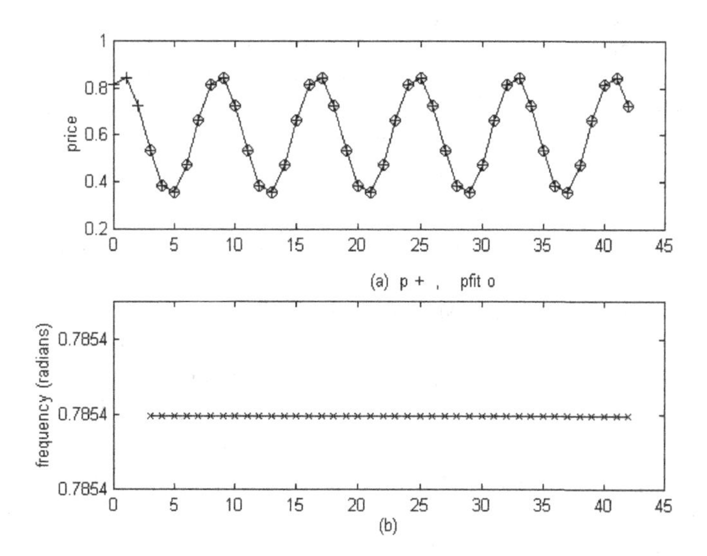

Fig 6.1(a) Input price data (marked as + and joined by a line) simulated as a sine wave plotted versus time, t. The calculated price data points (marked as o) are also plotted. They agree exactly with the input price data.
(b) Calculated frequencies (marked as x) are plotted. They agree exactly with the input frequency (drawn as a line).

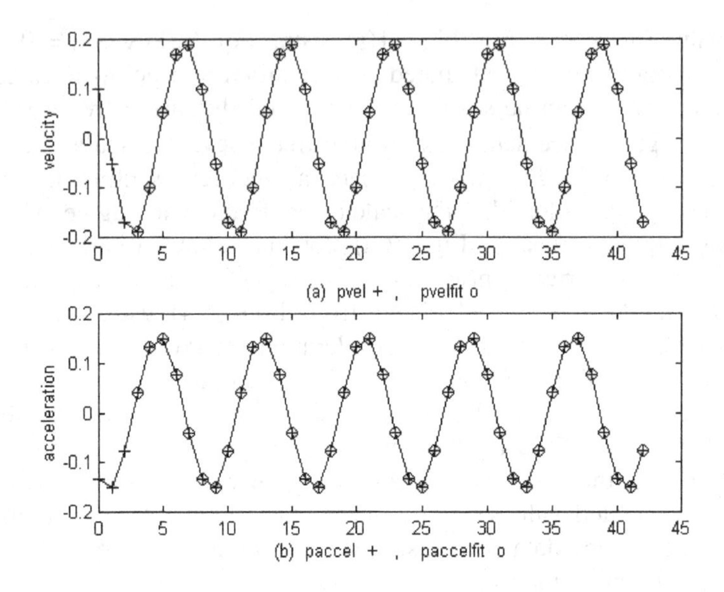

Fig 6.2(a) Wave velocity (marked as o), calculated using the solved parameters, are plotted. The velocity (marked as + and joined by a line), as derived from the original input price data, is also plotted for comparison.
(b) Wave acceleration (marked as o) calculated using the solved parameters, are plotted. The acceleration (marked as + and joined by a line) as derived from the original input price data, is also plotted for comparison.

6.5 Alternate Calculation of Frequency (5 data points)

Calculating the frequency using 4 data points can have a problem some of the time. The factor inside the square root sign in Eqn (6.2) can happen to be negative, or the argument of arsine can happen to lie outside the range of –1 to 1, implying that the data cannot be modeled as a sine wave imposed on a constant level. Furthermore, x_{-1} - x_{-2} in Eqn (6.2) can equal to zero, or approximately equal to zero, causing ω to be undefined or yielding a large error. In these cases, ω cannot or should not be calculated from Eqn (6.2). When this happens, five data points could be used instead of four. As a matter of fact, four data points are still actually used, as x_{-2} is not employed in the calculation. The circular frequency, ω, would be given by:

$$\omega = \cos^{-1}\left[\frac{1}{2}\left(\frac{x_0 - x_{-4}}{x_{-1} - x_{-3}}\right)\right] \tag{6.7}$$

where x_{-4} is the closing price of four bars ago.

Eqn (6.7) implies that ω or T can be calculated instantly, and the time delay is only 2.5 bars, which is half the number of data points. In Eqn (6.7), the argument of arccosine can lie outside the range of –1 to 1, implying that the data cannot be modeled as a sine wave imposed on a constant level. In these cases, ω cannot be calculated from Eqn (6.7). Furthermore, x_{-1} - x_{-3} can equal to zero or approximately equal to zero, causing ω to be undefined or yielding a large error. We can calculate the errors of ω from both Eqn (6.2) and (6.7) (see Appendix 3). The ω which yields the lesser error would be chosen.

6.6 Example with a Frequency Chirp

The method with 4 or 5 data points is tested on a sine wave with a frequency chirp, which can be written as

$$x = A \sin \phi_1 + D = A \sin[\ \omega_0 (1+ct)\ t + \phi] + D \tag{6.8}$$

where ϕ_1 is the phase
ω_0 is a fixed circular frequency
c is a constant.

As the circular frequency ω equals to the derivative of ϕ_1 with respect to t, ω is given by

$$\omega = d\phi_1/dt = \omega_0 (1+2ct) \tag{6.9}$$

A signal with $A = 0.25$, $\omega_0 = \pi/4$, $c = 0.1$, $\phi = 0$ and $D = 0.3$ is plotted as + in Fig 6.3(a). The frequency ω is calculated from Eqn (6.2) and (6.7). The ω with the lesser error is chosen (see Appendix 3). The parameters A, ϕ and D are then calculated and substituted in Eqn (6.1) to find x, which is plotted in Fig 6.3(a) as 'o'. It can be seen that the calculated values agree with the data (marked as '+' and joined by a line) quite well. The calculated ω (in 'x' and joined by a line) is plotted in Fig 6.3(b) and is compared with the theoretical ω (plotted as a straight line) calculated from the raw data using Eqn (6.9). It can be seen that the calculated ω has a lag of about 2 to 2½ data points (bars). This is, of course, caused by our using four or five data points to calculate ω. Other than that, the calculated ω's agree reasonably well with the theoretical ones.

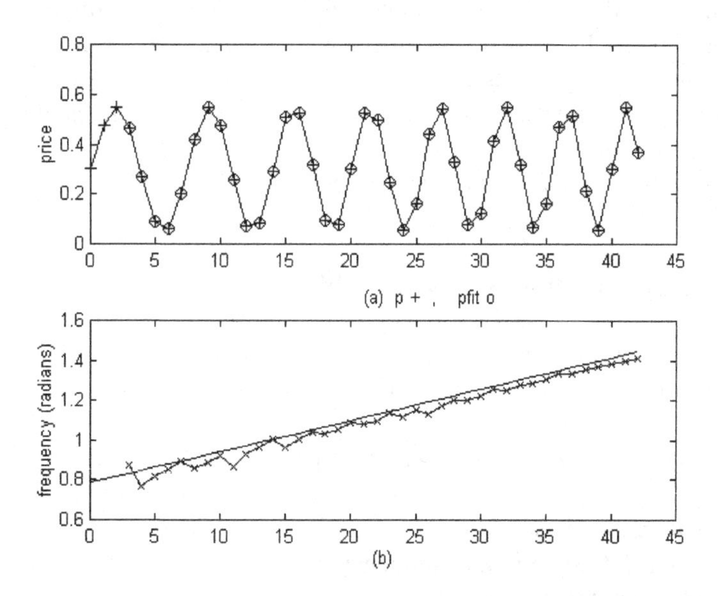

Fig 6.3(a) The calculated price, x, is plotted as 'o'. The calculated values agree with the original data (marked as '+' and joined by a line) quite well.
(b) The calculated ω (in x and joined by a line) is plotted and is compared with the theoretical ω (plotted as a straight line) calculated from the raw data using Eqn (6.9).

Fig 6.4(a) plots the velocity (marked as +) as derived from taking the slope (first derivative) of the original price data from Eqn (6.8), as well as the wave velocity (marked as o) calculated from Eqn (6.4) using the solved parameters. Fig 6.4(b) plots the acceleration (marked as +) as derived from taking the slope of the slope (second derivative) of the

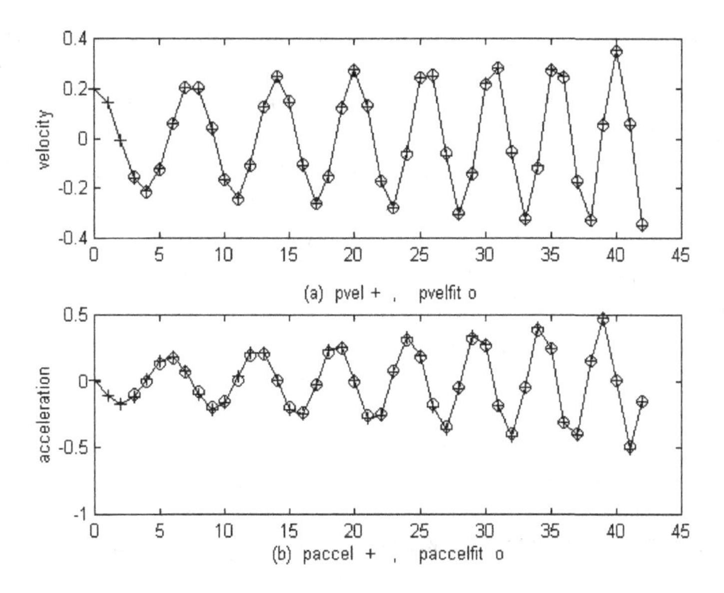

Fig 6.4(a) The wave velocity (marked as o) calculated from Eqn (6.4) using the solved parameters, is plotted together with the velocity (marked as + and joined by a line) as derived from taking the slope (first derivative) of the original price data from Eqn (6.8).
(b) The wave acceleration (marked as o) calculated from Eqn (6.6) using the solved parameters, is plotted together with the acceleration (marked as + and joined by a line) as derived from taking the slope of the slope (second derivative) of the original price data from Eqn (6.8).

original price data from Eqn (6.8), as well as the wave acceleration (marked as o) calculated from Eqn (6.6) using the solved parameters. Both the calculated wave velocity and acceleration values agree with the theoretically derived data quite well. It should be noted that the calculated values has practically little phase shift, which makes modeling the price data with a sine wave more accurate than modeling with a polynomial function.

6.7 Example with Real Financial Data

We will now attempt to model real financial data piecemeal with Eqn (6.1) and see how well the method works. The top plot in Fig 6.5

shows the daily S&P500 data in Japanese Candlesticks smoothed with an adaptive moving average (ama) with a smoothness factor of 32 (shown as a line). The adaptive moving average constucted by Jurik research is employed here. The bottom (i.e., the fourth) plot shows the circular frequency, ω, calculated from four or five smoothed data points. Some of the frequencies are not calculated because the smoothed financial data points cannot be modeled as a sine wave superimposed on a constant level. The period, T, in days, can be calculated from $T = 2\pi/\omega$. The second plot shows the velocity (shown in dots) calculated from ω and other parameters using Eqn (6.4). They agree very well with the velocity (shown as a line) calculated using the cubic velocity indicator. The third plot shows the acceleration (shown in dots) calculated from ω and other parameters using Eqn (6.6). They agree reasonably with the acceleration (shown as a line) calculated using the cubic acceleration indicator.

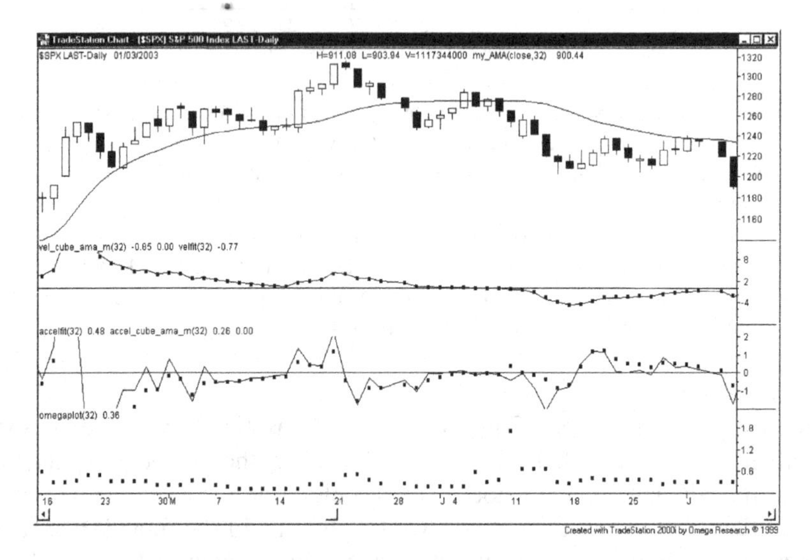

Fig 6.5 The top plot shows the daily S&P500 data in Japanese Candlesticks smoothed with an adaptive moving average (ama) with a smoothness factor of 32 (shown as a line). The bottom (i.e., the fourth) plot shows the circular frequency, ω, calculated from four or five smoothed data points. The second plot shows the velocity (shown in dots) calculated from ω and other parameters using Eqn (6.4). They agree very well with the velocity (shown as a line) calculated using the cubic velocity indicator. The third plot shows the acceleration (shown in dots) calculated from ω and other parameters using Eqn (6.6). *Chart produced with Omega Research TradeStation 2000i.*

The top plot in Fig 6.6 shows the same daily S&P500 data in Japanese Candlesticks but smoothed with an adaptive moving average (ama) with a smoothness factor of 3 (shown as a line). The bottom plot shows the circular frequency, ω, calculated from four or five smoothed data points. Again, some of the frequencies are not calculated because the smoothed financial data points cannot be modeled as a sine wave superimposed on a constant level. The period, T, in days, can be calculated from $T = 2\pi/\omega$. The second plot shows the velocity (shown in dots) calculated from ω and other parameters using Eqn (6.4). They agree reasonably well with the velocity (shown as a line) calculated using the cubic velocity indicator. The third plot shows the acceleration

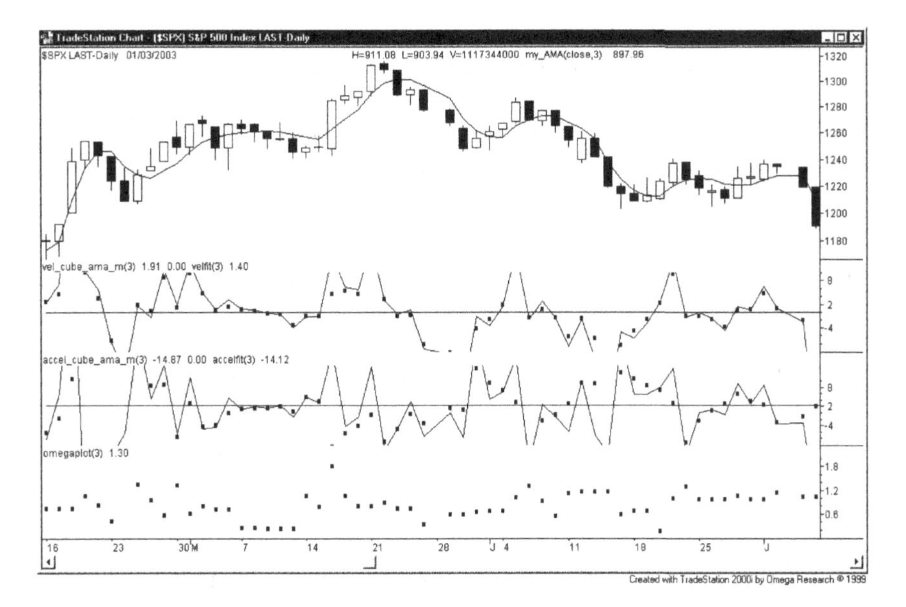

Fig 6.6 The top plot shows the daily S&P500 data in Japanese Candlesticks smoothed with an adaptive moving average (ama) with a smoothness factor of 3 (shown as a line). The bottom (i.e., the fourth) plot shows the circular frequency, ω, calculated from four or five smoothed data points. The second plot shows the velocity (shown in dots) calculated from ω and other parameters using Eqn (6.4). They agree reasonably well with the velocity (shown as a line) calculated using the cubic velocity indicator. The third plot shows the acceleration (shown in dots) calculated from ω and other parameters using Eqn (6.6). *Chart produced with Omega Research TradeStation 2000i.*

(shown in dots) calculated from ω and other parameters using Eqn (6.6). They agree reasonably with the acceleration (shown as a line) calculated using the cubic acceleration indicator. The acceleration agreements are not as good as the velocity agreements in both Fig 6.5 and 6.6. This is simply because modeling piecewise financial data by a sine wave is not perfect, causing error in calculating the circular frequency ω. This error propagates to calculating the velocity, which is the slope of the financial data. The error compounds even more when acceleration, the slope of the slope of the financial data, is calculated. It can be commented that when the sine wave velocity agrees with the cubic velocity, and the sine wave acceleration agrees with the cubic acceleration, the circular frequency ω calculated is more reliable. The adaptive moving average of smoothness 3 (ama3) is very similar to the exponential moving average with length 6 (ema6), and thus has approximately a time lag of one data point (bar) (see Fig 3.1(c)). The circular frequency ω calculated employs 4 or 5 data points, and thus has a time lag of 2 or 2½ data points (bars). As ω is calculated on the smoothed line using ama3, the total time lag is 3 or 3½ data points (bars). The time lag is thus much smaller than the lag of 20.5 data points (bars) in a recently proposed method of calculating frequency or period [Ehlers 2001]. However, because of different modeling techniques, the frequency calculated here may not be the same as those calculated at Ehlers [2001].

Computer programs, written in the EasyLanguage code of Omega Research's TradeStaion2000i, for calculating the frequency, wave velocity and wave acceleration are listed in Appendices A3.5 and A3.6.

6.8 Example with Real Financial Data (more stringent condition)

A more stringent condition would be if either the 4 *or* 5 data points cannot fit the sine wave model, the error of omega would be arbitrarily set to 20. The circular frequency, ω is then given an arbitrary negative number, so that it would not be plotted within the range specified (Fig 6.7 and 6.8). Fig 6.7 is essentially the same plot as Fig 6.5, except that the more stringent condition is applied. Again, Fig 6.8 is essentially the same plot as Fig 6.6, except that the more stringent condition is applied. It should be noted that Fig 6.7 and 6.8 have less frequencies plotted than in Fig 6.5 and 6.6. As well, many of the points where there are discrepancies between wave acceleration indicators and the cubic acceleration indicators in Fig 6.5 and 6.6 are eliminated in Fig 6.7 and 6.8.

Fig 6.7 The top plot shows the daily S&P500 data in Japanese Candlesticks smoothed with an adaptive moving average (ama) with a smoothness factor of 32 (shown as a line). The bottom (i.e., the fourth) plot shows the circular frequency, ω, calculated from four or five smoothed data points. The more stringent condition is applied to choose ω The second plot shows the velocity (shown in dots) calculated from ω and other parameters using Eqn (6.4). They agree very well with the velocity (shown as a line) calculated using the cubic velocity indicator. The third plot shows the acceleration (shown in dots) calculated from ω and other parameters using Eqn (6.6). *Chart produced with Omega Research TradeStation 2000i.*

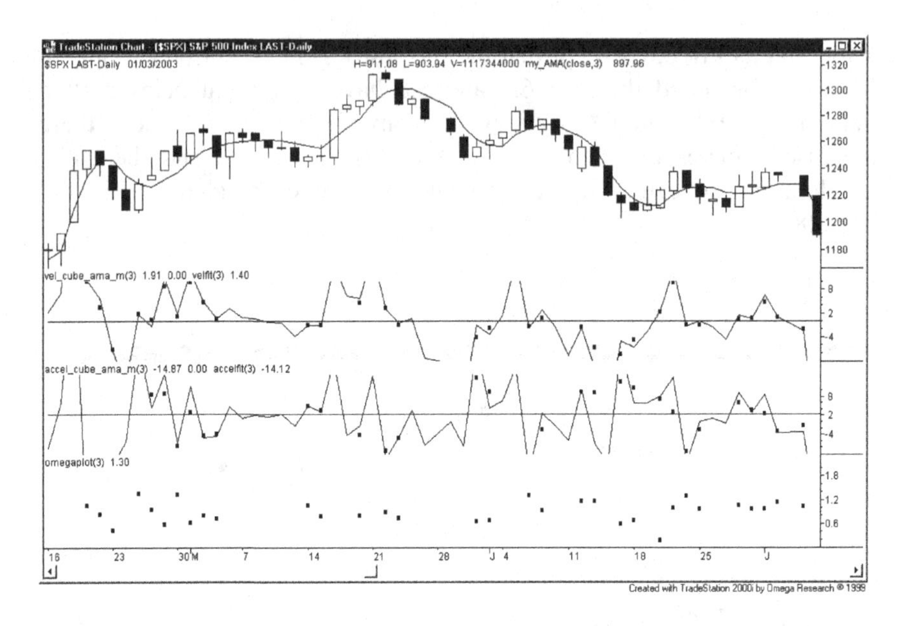

Fig 6.8 The top plot shows the daily S&P500 data in Japanese Candlesticks smoothed with an adaptive moving average (ama) with a smoothness factor of 3 (shown as a line). The bottom (i.e., the fourth) plot shows the circular frequency, ω, calculated from four or five smoothed data points. The more stringent condition is applied to choose ω. The second plot shows the velocity (shown in dots) calculated from ω and other parameters using Eqn (6.4). They agree quite well with the velocity (shown as a line) calculated using the cubic velocity indicator. The third plot shows the acceleration (shown in dots) calculated from ω and other parameters using Eqn (6.6). *Chart produced with Omega Research TradeStation 2000i.*

Chapter 7

Phase

Traders depend on indicators to tell them where the market is heading. For example, trends can be identified by trending indicators, which are actually causal low pass filters with phase or time lag. As traders would like to know changing market movement as early as possible, reducing phase lag would be of particular interest. One approach to reduce the phase lag has been discussed in Chapter 4. Another approach is to ask whether it is possible to predetermine the phase with respect to the frequency range, and work backward to find out what the indicator should be? We will develop a method in this chapter to solve this problem - for limited cases.

In signal processing, a system is an operator or a mapping that transform an input signal into an output signal by means of a fixed set of operations [Mak 2003]. A system is causal if the output of the system at any time depends only on present and past inputs, but not on future inputs. A system which is not causal is noncausal. A noncausal system has an output which depends on present and past inputs, as well as on future inputs. Thus, in real-time signal processing, as future values of the signal cannot be observed, a noncausal system is physically unrealizable.

In trading the financial market, as no future value is available, only causal system can be implemented. Causality implies a strong relationship between $H_R(\omega)$ and $H_I(\omega)$, the real and imaginary components of the frequency response $H(\omega)$ of a system. This relationship is discussed in the next section.

7.1 Relation between the Real and Imaginary Parts of the Fourier Transform of a Causal System

The relationship between the real and imaginary components of the Fourier Transform of a causal system is given by [Proakis and Manolaskis 1996] :

$$H_I(\omega) = -\frac{1}{2\pi}\int_{-\pi}^{\pi} H_R(\lambda)\cot\frac{\omega-\lambda}{2}\,d\lambda \tag{7.1}$$

Thus, $H_I(\omega)$ is uniquely determined from $H_R(\omega)$ through Eqn (7.1). The integral in Eqn (7.1) is called a discrete Hilbert Transform. As an example, we can take a look at the two point moving average, whose coefficients are given by (½, ½). The Fourier transform of the two point moving average is given by [Strang 1997, Mak 2003]:

$$H(\omega) = \tfrac{1}{2} + \tfrac{1}{2}\exp(-i\omega) = \cos^2(\omega/2) - i\,\tfrac{1}{2}\sin(\omega) \tag{7.2}$$

Substituting the real part of $H(\omega)$ into Eqn (7.1), we have

$$H_I(\omega) = -\frac{1}{2\pi}\int_{-\pi}^{\pi} \cos^2\frac{\lambda}{2}\cot\frac{\omega-\lambda}{2}\,d\lambda \tag{7.3}$$

Writing $\lambda' = (\omega - \lambda)/2$, Eqn (7.3) can be written as

$$H_I(\omega) = \frac{1}{\pi}\int_{(\omega+\pi)/2}^{(\omega-\pi)/2} \cos^2\left(\frac{\omega}{2}-\lambda'\right)\cot\lambda'\,d\lambda' \tag{7.4}$$

After simplifying, and using the fact that

$$\int_{(\omega+\pi)/2}^{(\omega-\pi)/2} \cot\lambda'\,d\lambda' = 0 \tag{7.5}$$

we get

$$H_I(\omega) = -\tfrac{1}{2}\sin(\omega) \tag{7.6}$$

which is the imaginary part of Eqn (7.2)

7.2 Calculation of the Frequency Response Function, H(ω)

Eqn (7.1) implies that, for a causal system, if the phase $\phi(\omega)$ is given, it may be possible to calculate $H(\omega)$. The phase $\phi(\omega)$ is given by

$$\tan\phi(\omega) = H_I(\omega)/H_R(\omega) \tag{7.7}$$

Substituting Eqn (7.7) into Eqn (7.1) yields

$$H_R(\omega)\tan\phi(\omega) = -\frac{1}{2\pi}\int_{-\pi}^{\pi} H_R(\lambda)\cot\frac{\omega-\lambda}{2}\,d\lambda \tag{7.8}$$

In computer calculation, the integral in Eq(7.8) needs to be changed to a discrete summation with

$$-\pi = \lambda_0, \lambda_1, \lambda_2, \ldots\ldots, \lambda_n = \pi$$

If $\phi(\omega)$ is known or given, setting ω to be one of the λ_i's will yield n+1 equations, with n+1 unknowns $H_R(\lambda_i)$. However, it would mean that in each equation, one of the cotangent term would become infinity when ω equals the λ_i that it is set equal to. Thus, we need to set each of the ω, ω_i , to be slightly away from the λ_i's , and then make the approximation on the LHS of Eqn (7.8) that $H_R(\omega_i) \sim H_R(\lambda_i)$ so that the equations can be solved. In order to yield an accurate calculation of the integral, ω_i has to be set equal to $\lambda_i + \delta\lambda/2$, where $\delta\lambda = \lambda_i - \lambda_{i-1}$.

From Eqn (7.8), the n+1 equations can be written as follows:

$$r_0 t_0 = -(r_0 c_{00} + r_1 c_{01} + r_2 c_{02} + \ldots\ldots + r_n c_{0n})\,\delta\lambda/(2\pi)$$

$$r_1 t_1 = -(r_0 c_{10} + r_1 c_{11} + r_2 c_{12} + \ldots\ldots + r_n c_{1n})\,\delta\lambda/(2\pi)$$

$$r_2 t_2 = -(r_0 c_{20} + r_1 c_{21} + r_2 c_{22} + \ldots\ldots + r_n c_{2n})\,\delta\lambda/(2\pi)$$

....

$$r_n t_n = -(r_0 c_{n0} + r_1 c_{n1} + r_1 c_{n2} + \ldots\ldots + r_n c_{nn})\,\delta\lambda/(2\pi) \qquad (7.9)$$

where $r_i = H_R(\lambda_i) \sim H_R(\omega_i)$

$t_i = \tan\phi\,(\omega_i)$

$c_{ij} = \cot[(\omega_i - \lambda_j)/2]$

One of the $H_R(\lambda_i)$ can be set arbitrarily. Since the magnitude of $H(\omega)$ is usually equal to 1, and we would most likely prefer $\phi\,(\omega = 0) = 0$, we can set $r_m = H_R(\lambda_m = 0) = 1$, where $m = n/2 + 1$. The n+1 equations in (7.9) can be written as

$$-r_m c_{0m} = (2\pi r_0 t_0/\delta\lambda + r_0 c_{00}) + r_1 c_{01} + r_2 c_{02} + \ldots\ldots + r_n c_{0n}$$

$$-r_m c_{1m} = r_0 c_{10} + (2\pi r_1 t_1/\delta\lambda + r_1 c_{11}) + r_2 c_{12} + \ldots\ldots + r_n c_{1n}$$

$$-r_m c_{2m} = r_0 c_{20} + r_1 c_{21} + (2\pi r_2 t_2/\delta\lambda + r_2 c_{22}) + \ldots\ldots + r_n c_{2n}$$

....

$$-r_m c_{nm} = r_0 c_{n0} + r_1 c_{n1} + r_1 c_{n2} + \ldots\ldots + (2\pi r_n t_n/\delta\lambda + r_n c_{nn}) \qquad (7.10)$$

The LHS of Eqn (7.10) can be set as constants as described above. That leaves n unknowns, r_i's , $i \neq m$, in n+1 equations. For this over-determined case of having more equations than unknowns, a least square solutions of the unknowns can be found [Hanselman and Littlefield 1997].

After $r_i = H_R$ are found, H_I can be determined from Eqn (7.1). The filter H is then completely determined.

7.2.1 *Example — The Two Point Moving Average*

We will use the two point moving average as an example. We know that the phase of its $H(\omega)$, $\phi(\omega)$, is given by [Mak 2003]:

$$\phi(\omega) = -\omega/2 \tag{7.11}$$

Knowing only this phase, we will reconstruct $H(\omega)$ using the above method. The real part of $H(\omega)$ is shown in Fig 7.1. The calculated values are compared with the theoretical values which are given by the real part of the RHS of Eqn (7.2). The slight discrepancy is caused by the approximation as described above.

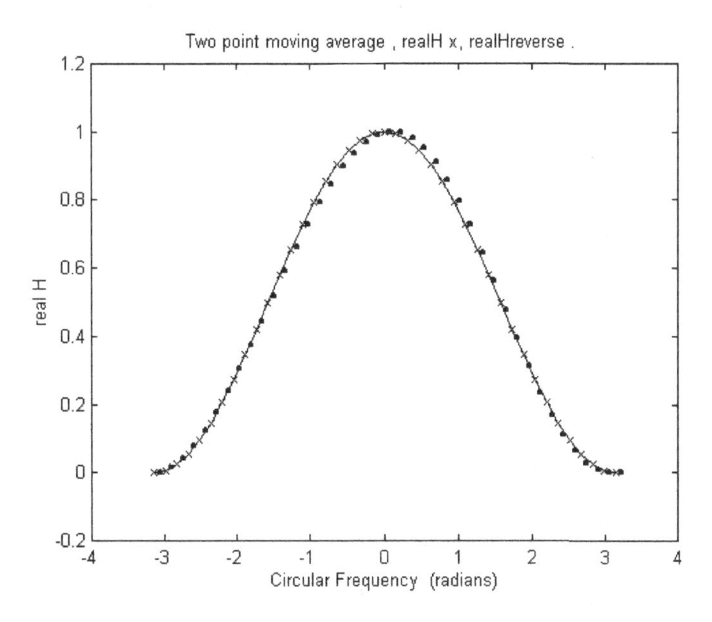

Fig 7.1 The real part of the Fourier Transform of the two point moving average are calculated using the method described in this chapter. The calculated values (plotted as .) are compared with the theoretical values (plotted as x and joined by a line) which are given by the real part of the RHS of Eqn (7.2).

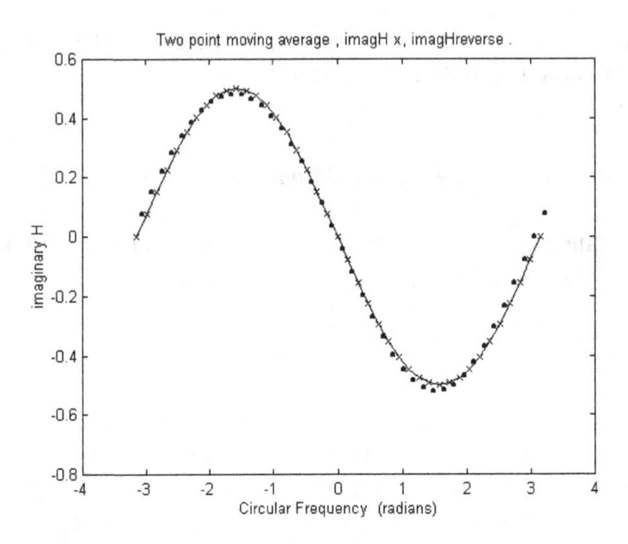

Fig 7.2 The imaginary part of the Fourier Transform of the two point moving average are calculated using the method described in this chapter. The calculated values (plotted as .) are compared with the theoretical values (plotted as x and joined by a line) which are given by the imaginary part of the RHS of Eqn (7.2).

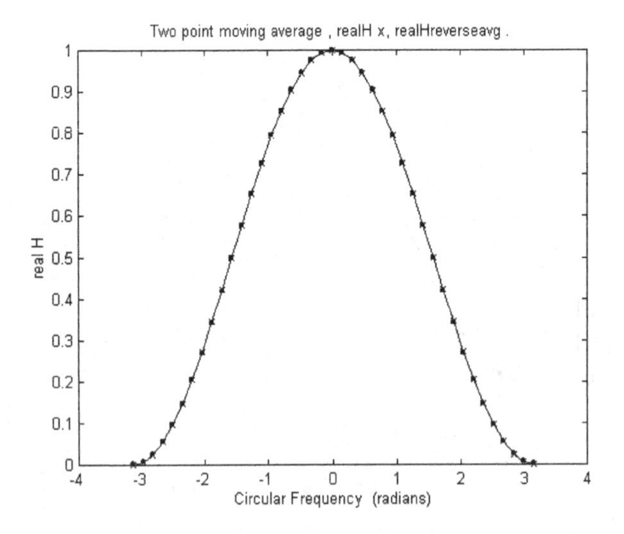

Fig 7.3 The real part of the Fourier Transform of the two point moving average are calculated using the improved method. The calculated values (plotted as .) are compared with the theoretical values (plotted as x and joined by a line) which are given by the real part of the RHS of Eqn (7.2). The agreement is much better than that in Fig 7.1.

The imaginary part of $H(\omega)$ is shown in Fig 7.2 . The calculated values are compared with the theoretical values which are given by the imaginary part of the RHS of Eqn (7.2). Again, the slight discrepancy is caused by the approximation as described above.

The discrepancy can be decreased by calculating the integral in Eqn (7.8) by setting ω_i to be equal to $\lambda_i - \delta\lambda/2$, where $\delta\lambda = \lambda_i - \lambda_{i-1}$. The $H_R(\omega_i)$ calculated will be averaged with the $H_R(\omega_i)$ calculated earlier when ω_i was set equal to $\lambda_i + \delta\lambda/2$.

The average $H_R(\omega_i)$ are plotted in Fig 7.3 and compared with the theoretical values which are given by the real part of Eqn (7.2). The agreement is much better than that in Fig 7.1. The average $H_I(\omega_i)$ are plotted in Fig 7.4 and compared with the theoretical values which are given by the imaginary part of Eqn (7.2). The agreement is much better than that in Fig 7.2. The average $H_R(\omega_i)$ and the average $H_I(\omega_i)$ can be used to calculate the average magnitude of $H(\omega_i)$ and compared with the theoretical values calculated in Eqn (7.2). These are plotted in Fig 7.5. The phase calculated from the average $H_R(\omega_i)$ and the average $H_I(\omega_i)$ is plotted in Fig 7.6, and compared with the theoretical phase in Eqn (7.11). The agreement is good except for the two end points. The discrepancy is caused by the very small disagreement between the calculated average of the real and imaginary parts with the corresponding theoretical values.

The unit impulse response of a system, $h(k)$ is related to $H(\omega)$ by the Discrete Time Fourier Transform [Oppenheim et al 1999].

$$h(k) = \frac{1}{2\pi}\int_{-\pi}^{\pi} H(\omega)\exp(i\omega k)d\omega \tag{7.12}$$

From Eqn (7.12) and using the average $H_R(\omega_i)$ and the average $H_I(\omega_i)$, $h(k)$ can be calculated. They are plotted in Fig 7.7, with $h(0) = 0.5000$, $h(1) = 0.4993$, and $h(k) \sim 0$ for $k > 1$. This result agrees very well with the theoretical values, $h(0) = h(1) = ½$ in Eqn (7.2).

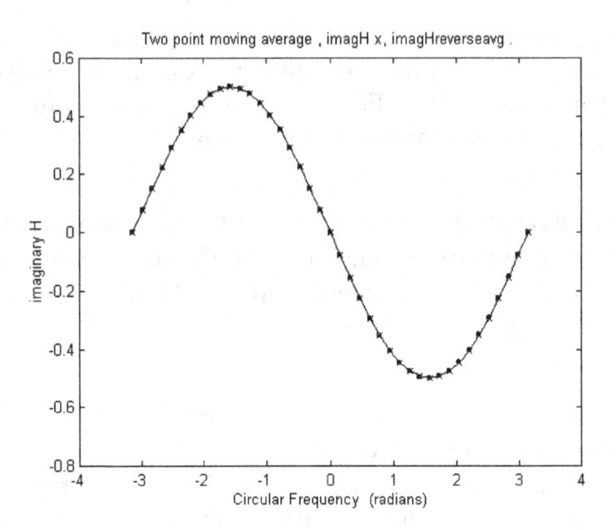

Fig 7.4 The imaginary part of the Fourier Transform of the two point moving average are calculated using the improved method. The calculated values (plotted as .) are compared with the theoretical values (plotted as x and joined by a line) which are given by the imaginary part of the RHS of Eqn (7.2). The agreement is much better than that in Fig 7.2.

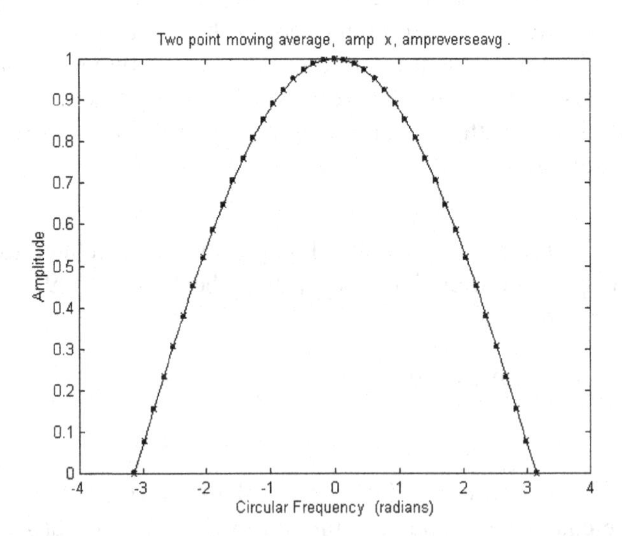

Fig 7.5 The magnitude of the Fourier Transform of the two point moving average are calculated using the improved method. The calculated values (plotted as .) are compared with the theoretical magnitude (plotted as x and joined by a line) which can be calculated from the RHS of Eqn (7.2).

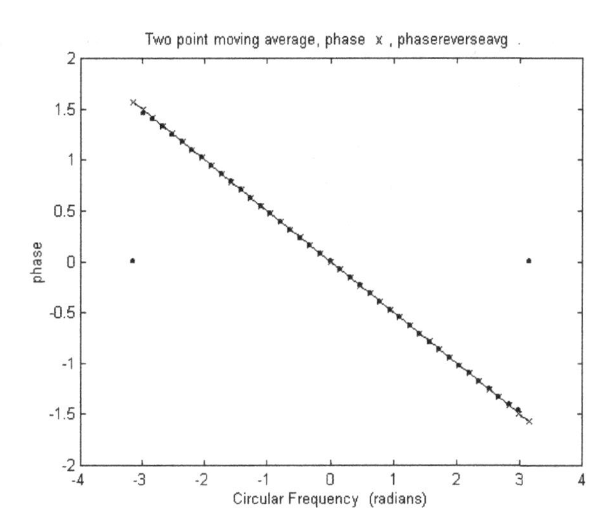

Fig 7.6 The phase of the Fourier Transform of the two point moving average are calculated using the improved method. The calculated values (plotted as .) are compared with the theoretical phase (plotted as x and joined by a line) given by Eqn (7.11). The agreement is good except for the two end points. The discrepancy is caused by the very small disagreement between the calculated averages of the real and imaginary parts with the corresponding theoretical values.

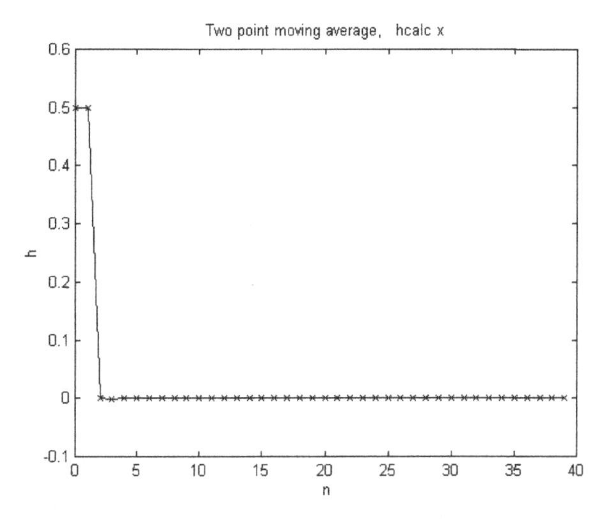

Fig 7.7 The unit impulse response of the two point moving average are calculated using the improved method. The calculated values (plotted as x and joined by a line) agree very well with the theoretical values, which are given as h(0) = h(1) = ½ in Eqn (7.2).

7.3 Computer Program for Calculating H(ω) and h(n) of a Causal System

The computer program for calculating H(ω) and h(n), given only the phase, ϕ(ω), has been written in MATLAB programming language, and is listed below:

```
% phasegivenprog1book,
% Given only the phase of the Fourier Transform (FT)
of the two point moving average, H, calculate the
real and imaginary part of H
% use both angle0 and angle1 to calculate the
integral; take average to calculate h
clear
mend=40; % mend = n in book
m=(0:1:mend);
nend=mend - 1; % used for h later
n=(0:1:nend);
dang = 2*pi/mend;% interval for integration
ang=-pi + dang*m % ang ranges from -pi to pi
angle0=ang + dang/2;% shifted from ang to avoid
infinity in cotangent
angle1=ang - dang/2;% shifted from ang to avoid
infinity in cotangent
% Set up the theoretical values for the Fourier
Transform of the two point moving average
% These theoretical values are used for comparing
with the calculated values later.  They are not used
for calculations.
for L=1:mend+1
    H(L)= 0.5 + 0.5*exp(-i*ang(L));% FT of two point
moving average
    amp(L)=abs(H(L));
   phase(L)= -ang(L)/2;% This is also equal to
angle(H(L))
      realH(L)=real(H(L));
   imagH(L)=imag(H(L));
end
% End of set up
figure(1)% Plot theoretical value of phase of two
point moving average for illustration purpose
plot(ang,phase,'k.-')
```

```
title(' Two point moving average ')
xlabel('Circular Frequency  (radians)');
ylabel('Phase  (radians)' )
% Calculation starts here
rm =1; % set an arbitrary value to the element of
real H when omega = 0, usually 1
for I= 1:mend +1
     for J=1:mend +1
         r(I,J) = 1;
     end
end
for I=1:mend + 1
      r(I,mend/2+1)= rm; %set realH(omega=0) to be
rm, usually 1
end
for I=1:mend+1
   factor =2;% factor can be changed to other number,
e.g., 3,to calculate another moving average H
   phasegiven(I)= -angle0(I)/factor;
   phasegiven1(I)= -angle1(I)/factor;
    for J=1:mend+1
      if(I==J)
         C(I,J)=
r(I,J)*((2*pi/dang)*tan(phasegiven(I))+ cot(
(angle0(I) - ang(J))/2));% r(I,J) does not have to be
included, this is just a convenient way to set a
value to real H(omega = 0)
         C1(I,J)=
r(I,J)*((2*pi/dang)*tan(phasegiven1(I))+ cot(
(angle1(I) - ang(J))/2));
      else
         C(I,J)= r(I,J)*cot( (angle0(I) -
ang(J))/2);% r(I,J) does not have to be included,
this is just a convenient way to set a value to real
H(omega = 0)
         C1(I,J)= r(I,J)*cot( (angle1(I) -
ang(J))/2);
      end
    end
end
% Set up matrix equation
for I=1:mend+1
     bb(I)= -C(I,mend/2+1);
     bb1(I)= -C1(I,mend/2+1);
```

```
end
  BB = bb';% BB is the transpose of matrix bb
  BB1 = bb1'; % BB1 is the transpose of matrix bb1
for I=1:mend+1
     for J=1:mend/2
        AA(I,J)=C(I, J);
        AA1(I,J)=C1(I, J);
     end
end
for I=1:mend+1
     for J=mend/2+1:mend
        AA(I,J)=C(I, J+1);
        AA1(I,J)=C1(I, J+1);
     end
end
realHminus1 = AA\BB;% MATLAB P77
realHminus11 = AA1\BB1;%Solution of matrix equation
found
realHreverse(mend/2+1)= rm;
realHreverse1(mend/2+1)= rm;
for I=1:mend/2
   realHreverse(I)=realHminus1(I);
   realHreverse1(I)=realHminus11(I);
end
for I=mend/2+2:mend+1
   realHreverse(I)=realHminus1(I-1);
   realHreverse1(I)=realHminus11(I-1);
end
for I=1:mend+1
   realHreverseavg(I) = (realHreverse(I) +
realHreverse1(I))/2;
end
for M=1:mend+1
for L=1:mend+1
   yH(L) = realHreverseavg(L)*cot ((angle0(M) -
ang(L))/2);% Proakis and Manolakis 1996,  P618
   yH1(L) = realHreverseavg(L)*cot ((angle1(M) -
ang(L))/2);
end
imagHreverse(M)= -(1/(2*pi))*trapz(ang, yH);
imagHreverse1(M)= -(1/(2*pi))*trapz(ang, yH1);
end
for I=1:mend+1
```

```
   imagHreverseavg(I)= (imagHreverse(I) +
imagHreverse1(I))/2;
   ampreverseavg(I)= sqrt(realHreverseavg(I)^2+
imagHreverseavg(I)^2);
   phasereverseavg(I) =
atan(imagHreverseavg(I)/realHreverseavg(I));
end
figure(2)
plot(ang,realH,'bx-',ang,realHreverseavg, 'r.')
title('Two point moving average , realH x,
realHreverseavg . ')
xlabel('Circular Frequency  (radians)'); ylabel('real
H' )
figure(3)
plot(ang,imagH,'bx-',ang,imagHreverseavg, 'r.')
title('Two point moving average , imagH x,
imagHreverseavg . ')
xlabel('Circular Frequency  (radians)');
ylabel('imaginary H' )
figure(4)
plot(ang,amp, 'kx-', ang, ampreverseavg,'r.')
title(' Two point moving average,  amp  x,
ampreverseavg . ')
xlabel('Circular Frequency  (radians)');
ylabel('Amplitude' )
figure(5)
plot(ang,phase, 'kx-', ang, phasereverseavg, 'r.')
title(' Two point moving average, phase  x ,
phasereverseavg  .')
xlabel('Circular Frequency  (radians)');
ylabel('phase')
% Calculate h
for k=1:nend+1
   for L=1:mend+1
      integrand(L)=
(realHreverseavg(L)+i*imagHreverseavg(L))*
exp(i*ang(L)*(k-1));
   end
   hcalc(k)=(1/(2*pi))*trapz(ang, integrand);% Mak
2003, P146
end
figure(6)
plot(n, hcalc, 'kx-')
title(' Two point moving average,   hcalc x ')
xlabel('  n  '); ylabel('h')
```

7.3.1 *Example,* $\phi(\omega) = -\omega/3$

In the above program, if we change the phase given, we will get a different H(ω) and h(n). For example, if the phase is given as :

$$\phi(\omega) = -\omega/3 \tag{7.13}$$

the magnitude of H(ω) calculated would be different from that of the two point moving average. Fig 7.8 plots the magnitude of H(ω) and compared that with the magnitude of the two point moving average. Fig 7.9 plots the h(n) calculated.

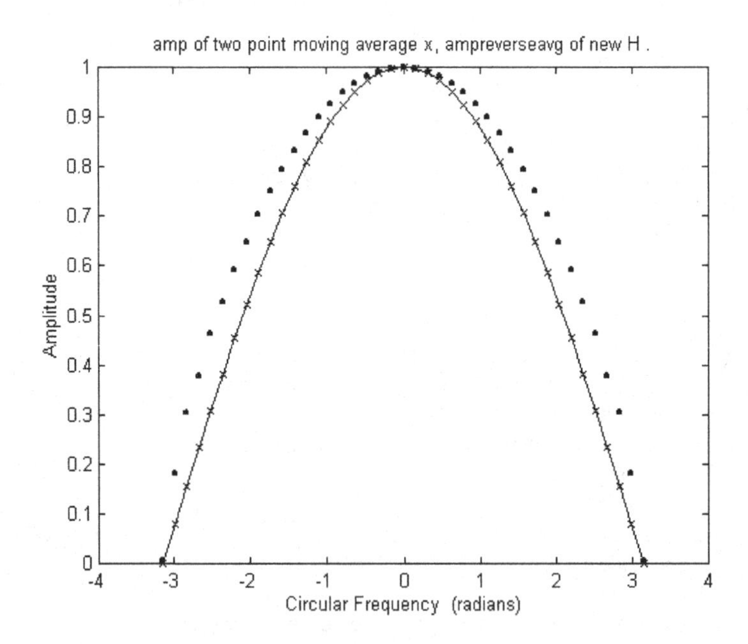

Fig 7.8 Given the phase, φ(ω) = -ω/3 , the magnitude of the Fourier Transform, |H(ω)| , is calculated using the improved method, and plotted as ' . ' . The magnitude of the Fourier Transform of the two point moving average is plotted for comparison (plotted as x and joined by a line).

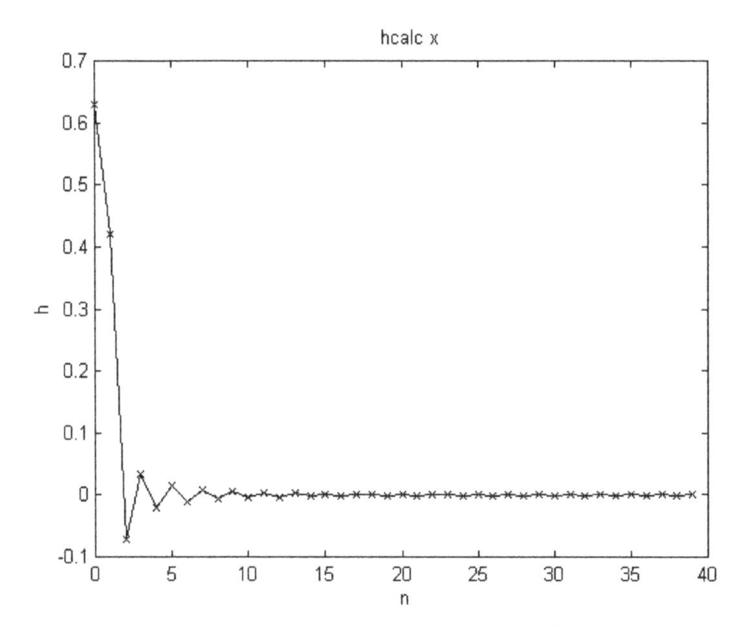

Fig 7.9 Given the phase, $\phi(\omega) = -\omega/3$, the unit impulse response, h(n) are calculated using the improved method. The calculated values are plotted as x and joined by a line.

7.3.2 *Example,* $\phi(\omega) = A\sin(\omega)$

In the computer program, the phase can be changed to another form. For example, it can be changed to:

$$\phi(\omega) = A\sin(\omega) \tag{7.14}$$

where A is the amplitude of the sine wave. For A = -0.8, the magnitude of H(ω) calculated is shown in Fig 7.10. The phase calculated is shown in Fig 7.11, and compared with the phase given in Eqn (7.14). The unit impulse response, h(n), calculated is shown in Fig 7.12. The form of the phase shown in Eqn (7.14) is somewhat similar to that of the exponential moving average (see Fig 3.1(b)). The magnitude of H(ω), and the unit impulse response, h(n), calculated are thus comparable to those of the exponential moving average.

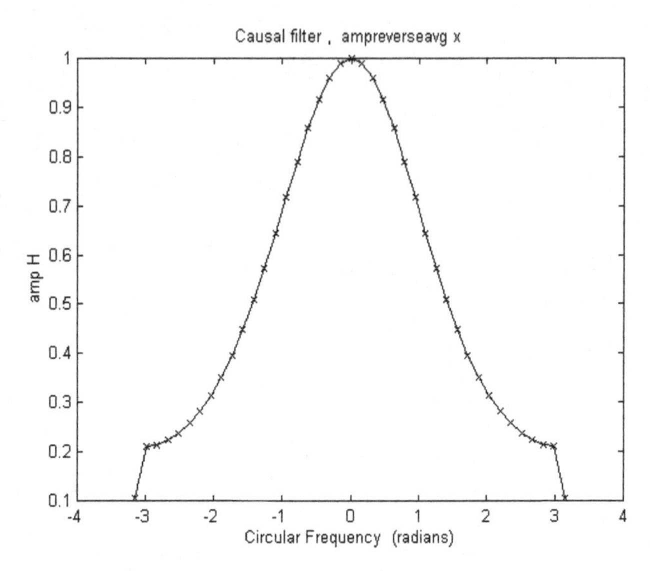

Fig 7.10 Given the phase, $\phi(\omega) = A\sin(\omega)$, the magnitude of the Fourier Transform, $|H(\omega)|$, is calculated using the improved method. It is plotted as x and joined by a line.

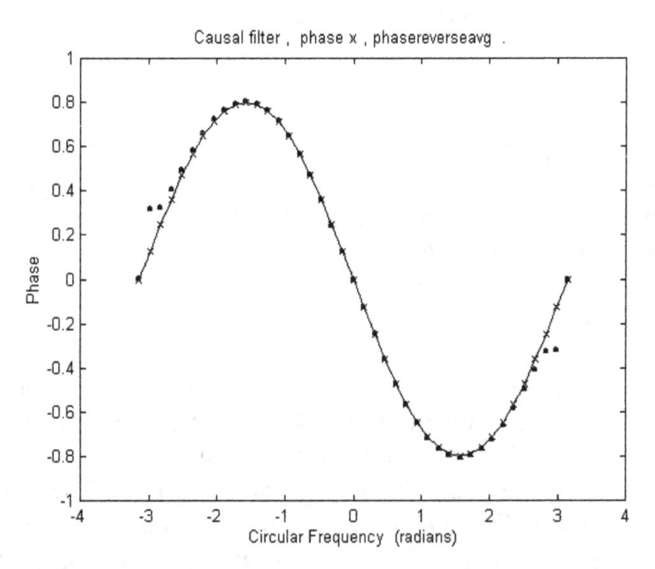

Fig 7.11 Given the phase, $\phi(\omega) = A\sin(\omega)$, the phase of the Fourier Transform is calculated using the improved method. The calculated values (plotted as .) are compared with the theoretical phase (plotted as x and joined by a line) given by Eqn (7.14).

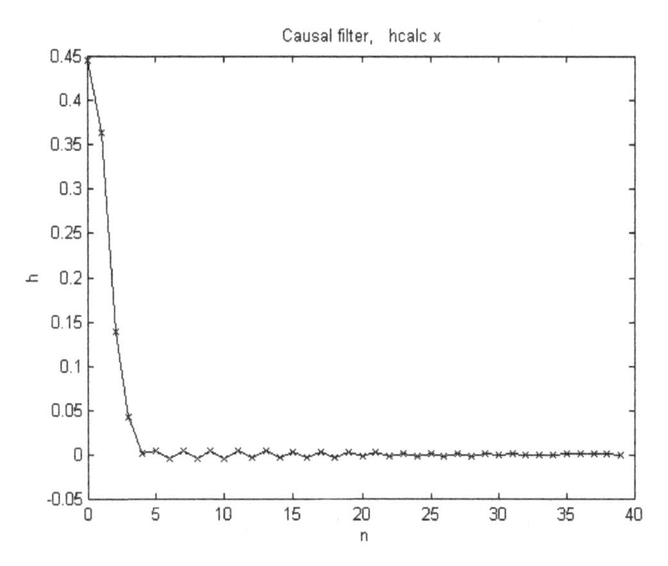

Fig 7.12 Given the phase, $\phi(\omega) = A\sin(\omega)$, the unit impulse response, h(n) are calculated using the improved method. The calculated values are plotted as x and joined by a line.

The mathematical technique described above can handle only some simple forms of phase spectrum as input. For other arbitrary phase spectrum inputs, other more robust mathematical techniques need to be developed.

7.4 Derivation of $H_R(\omega)$ in Terms of $H_I(\omega)$ for a Causal System

For completeness purpose, the relationship between $H_R(\omega)$ in terms of $H_I(\omega)$ is derived below:

The impulse response h(n) can be decomposed into an even and an odd sequence [Proakis and Manolaskis 1996]:

$$h(n) = h_e(n) + h_o(n) \tag{7.15}$$

where $h_e(n) = 1/2\,[h(n) + h(-n)]$ (7.16)

$$h_o(n) = 1/2\,[h(n) - h(-n)] \tag{7.17}$$

If h(n) is causal, it is possible to recover h(n) from its odd component $h_o(n)$ for $1 \leq n \leq \infty$. Since $h_o(n) = 0$ for n = 0, h(0) cannot be recovered from $h_o(n)$. We need to know h(0). It can be shown that

$$h(n) = 2\, h_o(n)\, u(n) + h(0)\, \delta(n) \qquad n \geq 0 \tag{7.18}$$

The Fourier transform for (7.18) is

$$H(\omega) = H_R(\omega) + jH_I(\omega)$$

$$= \frac{j}{\pi} \int_{-\pi}^{\pi} H_I(\lambda)\, U(\omega - \lambda)\, d\lambda \; + \; h(0) \tag{7.19}$$

where $U(\omega)$ is the Fourier transform of the unit step sequence u(n). Although u(n) is not absolutely summable, it has a Fourier transform [Proakis and Manolaskis 1996]

$$U(\omega) = \pi\delta(\omega) + \tfrac{1}{2} - \tfrac{1}{2}\, j \cot(\omega/2) \qquad -\pi \leq \omega \leq \pi \tag{7.20}$$

By substituting (7.20) into (7.19) and carrying out the integration, we obtain the relation between $H_I(\omega)$ and $H_R(\omega)$ as

$$H_R(\omega) \;=\; \frac{1}{2\pi} \int_{-\pi}^{\pi} H_I(\lambda) \cot\frac{\omega - \lambda}{2}\, d\lambda \; + \; h(0) \tag{7.21}$$

Chapter 8

Causal High Pass Filters

Traders quite often are interested in how fast the market is moving. Knowing the instantaneous frequency is one method to monitor the pace of the market. However, as discussed in Chapter 6, frequency can be difficult to be estimated accurately sometimes. An alternative method is to measure the slope or the first derivative (in Calculus) of the price data.

When the slope is positive, the market is heading up. When it increases, the market is heading up faster. When the slope is negative, the market is declining. When it decreases, the market is declining faster. When the slope is zero, it implies that the market is turning. Thus, the slope is an interesting parameter to be observed.

The market sometimes can turn down and then take a second wind to charge forward and go back up. It has been pointed out in Mak [2003] that the slope of the slope or the second derivative (in Calculus) of the price data may indicate whether the market is running out of gas. When the second derivative of a curve is positive, the curve is concaving up. When the second derivative of a curve is negative, the curve is concaving down [Protter and Morrey 1966]. Thus, if both the first and second derivatives are positive, the market is heading up. If the first derivative is positive but the second derivative is negative, it can imply the the market is running out of steam. Similarly, if both the first and second derivatives are negative, the market is heading down. If the first derivative is negative but the second derivative is positive, it can imply that the market would soon head back up. The first and second derivatives of the market price data can be named as the velocity and acceleration of the price. They are actually causal high pass filters.

Causal high pass filters affiliated with polynomials of degree 2 and 3 have been created in Mak [2003]. They are useful for monitoring the slope (or velocity) and slope of the slope (or acceleration) of the price data in trading. Here, we will first describe ideal filters for the slope and the slope of the slope. Then we will describe how causal velocity and acceleration filters can be created to simulate them. It will be pointed out that an old indicator popular with traders, momentum, is a linear (polynomial of degree one) velocity indicator. Velocity and acceleration indicators with polynomials of degree equal to and higher than 3 will be constructed. The phase response of the combination of these high pass filters with low pass filters will be detailed.

8.1 Ideal Filters

8.1.1 *The Slope*

A general signal can be written as

$$x(t) = \exp(i\omega t) \tag{8.1}$$

Taking the derivative of Eq (8.1), and changing the continuous time t to the integer n, we get an output y(n) given by

$$y(n) = i\omega\exp(i\omega n) = i\omega x(n) \tag{8.2}$$

The frequency response of an ideal digital differentiator, $H(\omega)$, can be thus written as [Strang 1997, Mak 2003]:

$$H(\omega) = y(n)/x(n) = i\omega = \omega\exp(i\pi/2) \tag{8.3}$$

The amplitude of the response is linearly proportional to the circular frequency ω, with a gradient equals to 1. $H(\omega)$ is thus a high pass filter. From Eq (8.3), it can be seen that the output signal has a phase lead of $\pi/2$ with respect to the input signal. Here, $H(\omega)$ is a non-causal filter and cannot be applied to real trading, where future data is not available. Causal velocity filters simulating this filter will be considered below. It should be noted that for a causal filter, the phase and amplitude are inter-related to each other, as discussed in Chapter 7.

8.1.2 *The Slope of the Slope*

Taking the second derivative of Eq (8.1), and changing the continuous time t to the integer n, we get an output y(n) given by

$$y(n) = -\omega^2 \exp(i\omega n) = -\omega^2 x(n) \quad (8.4)$$

The frequency response of an ideal digital second differentiator, $H(\omega)$, can be thus written as :

$$H(\omega) = -\omega^2 = \omega^2 \exp(i\pi) = \omega^2 \exp(-i\pi) \quad (8.5)$$

The amplitude of the response is proportional to the square of the circular frequency ω. $H(\omega)$ is thus a high pass filter. From Eq (8.5), it can be seen that the output signal has a phase lead (or phase lag) of π with respect to the input signal. Again, $H(\omega)$ here is a non-causal filter and cannot be applied to trading, where future data is not available. Causal acceleration filters simulating this filter will be considered below.

8.2 Momentum

8.2.1 *The Filter*

Momentum is an old and commonly used indicator employed in trading. It measures how fast the price changes [Elder 1993]. It has a different meaning from the term in Physics, where it is defined as mass times velocity. Momentum, in trading terminology, should be compared to the velocity term in Physics. It can be considered as a causal high pass filter affiliated with a polynomial of degree 1. The unit impulse response of a momentum indicator is defined as (1, 0, …., 0, 1). The output response, y, after the input price data, x, is filtered by the momentum indicator is

$$y(n) = x(n) - x(n-(N-1)) \qquad N \geq 2 \quad (8.6)$$

where x(n) is the closing price or the smoothed closing price of the nth bar

$x(n-(N-1))$ is the closing price or the smoothed closing price of (N-1) bar back from the nth bar

Thus, the momentum indicator can be considered as an N-point moving difference. The frequency response function of the momentum indicator $H(\omega)$, can be written as

$$H(\omega) = 1 - \exp[-i(N-1)\omega]$$

$$= 2\sin[(N-1)\omega/2]\exp\{i[\pi/2-(N-1)\omega/2]\} \quad (8.7)$$

Eq (8.7) means that the momentum indicator has a phase lag of $(N-1)\omega/2$ or an $(N-1)/2$ bars lag from an ideal phase lag of $\pi/2$ shown in Eq (8.3).

For N = 2, the output response of the 2-bar momentum indicator is given by

$$y(n) = x(n) - x(n-1) \quad (8.8)$$

From Eq (8.7), the 2-bar momentum indicator (or the 2-point moving difference) has a phase lag of $\omega/2$ or a half bar lag from an ideal phase lag of $\pi/2$. Its amplitude and phase response have been plotted in Mak [2003]. The phase will lag even more if the raw price data is first smoothed with a moving average before the momentum indicator is applied, as we will see in the next section.

8.2.2 *Filtering Smoothed Data*

To avoid the jumpiness of the momentum of price data, price data are quite often smoothed by a moving average (e.g., an exponential moving average) before a momentum indicator is applied [Elder 1993, Pring 1991]. Some traders apply the momentum first and then the moving average after. It is believed that the order of applying these two indicators would make a difference [Elder 1993]. It does not. This is because convolution is commutative [Hayes 1999], i.e.,

$$h_1(n)*h_2(n) = h_2(n)*h_1(n) \quad (8.9)$$

where $h_1(n)$ and $h_2(n)$ are the unit impulse response of the two indicators. Eq (8.9) would apply providing that the indicator is not adaptive, as the unit sample response of an adaptive indicator (e.g., adaptive moving average) changes from data point to data point.

It would be useful to take a look at the Fourier Transform of the convolution of the unit impulse responses of two indicators [Brigham 1974]:

$$F\{h_1 * h_2\} = F\{h_1\}F\{h_2\}$$

$$= r_1 \exp(i\theta_1) r_2 \exp(i\theta_2) = r_1 r_2 \exp[i(\theta_1 + \theta_2)] \quad (8.10)$$

where F is the Fourier Transform

r_1 and r_2 are the magnitudes of the complex numbers, $F\{h_1\}$and $F\{h_2\}$ respectively,

θ_1 and θ_2 are the phases of the complex numbers, $F\{h_1\}$and $F\{h_2\}$ respectively,

Thus, the magnitude and phase of $F\{h_1 * h_2\}$ is r_1 r_2 and $\theta_1 + \theta_2$ respectively.

We will take a look at an example. An exponential moving average of length 3 or 6 can be applied first to smooth the price data. The momentum indicator is then applied to the smoothed data. Fig 8.1(a) plots the amplitude responses of the momentum indicator, the combination of the momentum indicator with the exponential moving average of length 3, and the combination of the momentum indicator with the exponential moving average of length 6. The amplitude of the momentum is much decreased by the exponential moving average. However, this is not so critical as amplitude does not play a significant role in timing a trade.

Fig 8.1(b) plots the phase responses of the momentum indicator, the combination of the momentum indicator with the exponential moving average of length 3, and the combination of the momentum indicator with the exponential moving average of length 6. The lag in phase of the momentum indicator from the ideal phase of the slope is further increased by the exponential moving averages, rendering the combination not a good simulation of the slope. This is the reason why traders do not use the momentum as a slope. They use it as overbought and oversold indicators [Pring 1991, Mak 2003], which are somewhat arbitrary.

In order to improve on the phase lag, velocity indicator affiliated with cubic polynomial has been developed [Mak 2003]. This is further discussed in the next section.

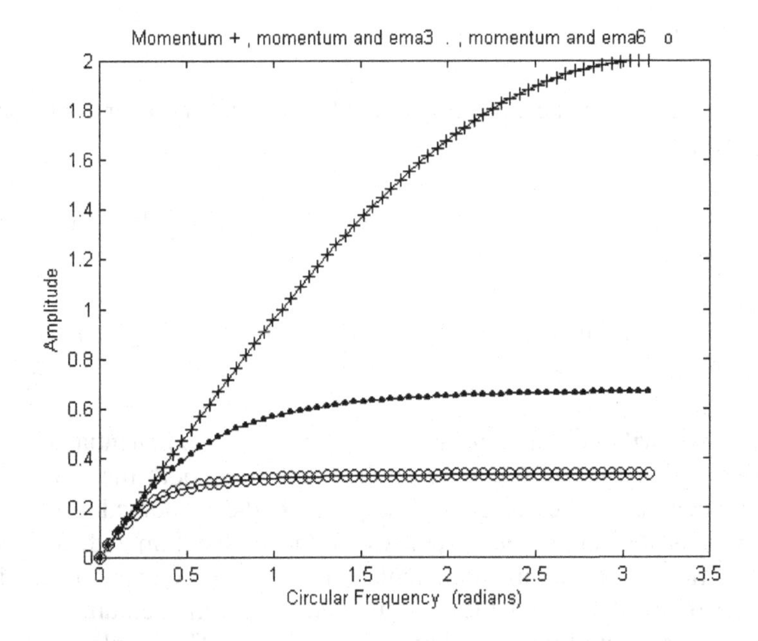

Fig 8.1(a) The amplitude responses of the momentum indicator (plotted as +), the combination of the momentum indicator with the exponential moving average of length 3 (plotted as .), and the combination of the momentum indicator with the exponential moving average of length 6 (plotted as o) are plotted versus ω, the circular frequency.

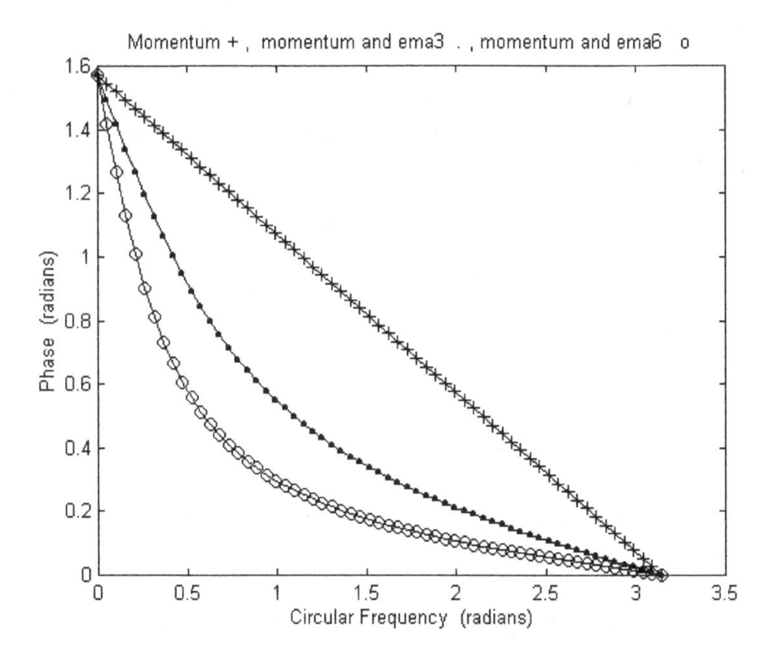

Fig 8.1(b) The phase responses of the momentum indicator (plotted as +), the combination of the momentum indicator with the exponential moving average of length 3 (plotted as .), and the combination of the momentum indicator with the exponential moving average of length 6 (plotted as o) are plotted versus ω, the circular frequency. At $\omega = 0$, the phase lead is $\pi/2$. At $\omega > 0$, the phase lags behind $\pi/2$.

8.3 Cubic Indicators

8.3.1 *The Filters*

The cubic velocity and acceleration indicators have been described in Mak [2003]. Basically, four adjacent market price data points are fitted to a cubic function. The slope and the slope of the slope of the cubic function at the most recent data point are then calculated using Calculus. They would represent the velocity and acceleration of the market price.

8.3.1.1 *Cubic Velocity Indicator*

The cubic velocity indicator is defined as (11/6, -3, 3/2, -1/3). The output response, y, after the input price data, x, is filtered by the cubic velocity indicator is

$$y(n) = \frac{11}{6}x(n) - 3x(n-1) + \frac{3}{2}x(n-2) - \frac{1}{3}x(n-3) \tag{8.11}$$

Therefore, the current velocity is given by

$$y(0) = \frac{11}{6}x(0) - 3x(-1) + \frac{3}{2}x(-2) - \frac{1}{3}x(-3) \tag{8.12}$$

where x(0) is the closing price or the smoothed closing price of the current bar

x(-1) is the closing price or the smoothed closing price of one bar ago

x(-2) is the closing price or the smoothed closing price of two bars ago

x(-3) is the closing price or the smoothed closing price of three bars ago

8.3.1.2 *Cubic Acceleration Indicator*

The cubic accelerator indicator is defined as (2, -5, 4, -1). The output response, y, after the input price data, x, is filtered by the cubic accelerator indicator is

$$y(n) = 2x(n) - 5x(n-1) + 4x(n-2) - x(n-3) \tag{8.13}$$

Therefore, the current acceleration is given by

$$y(0) = 2x(0) - 5x(-1) + 4x(-2) - x(-3) \tag{8.14}$$

8.3.2 *Filtering Smoothed Data*

As discussed above, price data are quite often smoothed first to eliminate the high frequencies before other indicators are applied on them. We will employ the exponential moving averages of length 3 and 6 to smooth the price data below.

8.3.2.1 *Cubic Velocity Indicator*

The amplitude and phase of the Discrete Time Fourier Transform (DTFT) of the cubic velocity indicator are plotted in Fig 8.2(a) and (b) respectively. Up to a circular frequency, ω, which is approximately 1.5 radians, it can be seen that the amplitude is directly proportional to ω, with a gradient equals to 1 (Fig 8.2(a)), and the phase is approximately

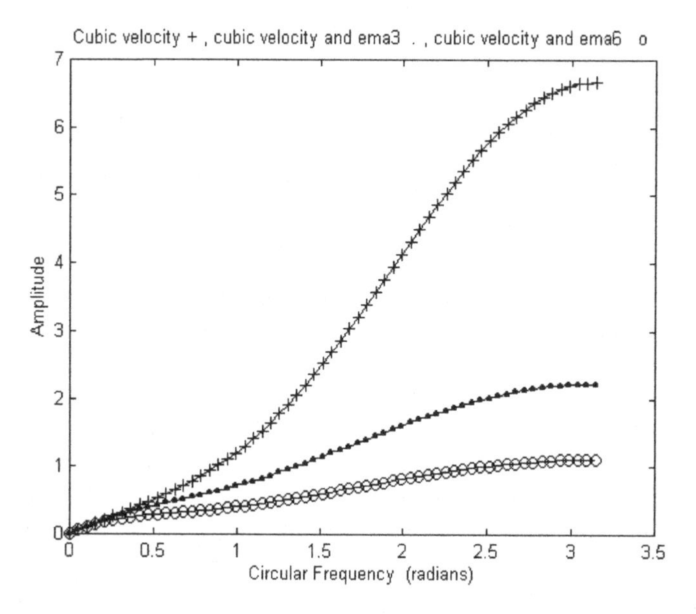

Fig 8.2(a) The amplitude responses of the cubic velocity indicator (plotted as +), the combination of the cubic velocity indicator with the exponential moving average of length 3 (plotted as .), and the combination of the cubic velocity indicator with the exponential moving average of length 6 (plotted as o) are plotted versus ω, the circular frequency.

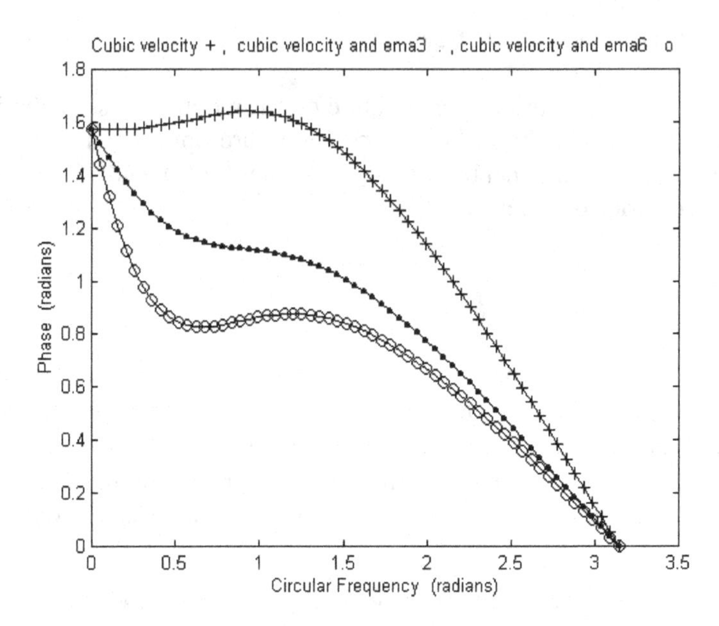

Fig 8.2(b) The phase responses of the cubic velocity indicator (plotted as +), the combination of the cubic velocity indicator with the exponential moving average of length 3 (plotted as .), and the combination of the cubic velocity indicator with the exponential moving average of length 6 (plotted as o) are plotted versus ω, the circular frequency.

$\pi/2$ (Fig 8.2(b)). This makes it a very good candidate to simulate the slope (cf Eqn (8.3)). The amplitude responses of the combination of the velocity indicator with the exponential moving average of length 3, and the combination of the velocity indicator with the exponential moving average of length 6 are also plotted in Fig 8.2(a). Furthermore, the phase responses of the combination of the velocity indicator with the exponential moving average of length 3, and the combination of the velocity indicator with the exponential moving average of length 6 are also plotted in Fig 8.2(b). Fig 8.2(b) shows that when the indicator is applied to the smoothed data, the combined phase lag makes it much less than the ideal phase lead of $\pi/2$. However, it should be noted that when the cubic velocity indicator is applied to data smoothed by some adaptive moving average, the combined phase lag is close to the ideal phase lead of $\pi/2$ (at least for part of the frequency range), thus making the cubic velocity indicator a good indicator to be employed.

8.3.2.2 *Cubic Acceleration Indicator*

The amplitude and phase of the Discrete Time Fourier Transform (DTFT) of the cubic acceleration indicator are plotted in Fig 8.3(a) and (b) respectively. Up to a circular frequency, ω, of approximately 0.5 radians, it can be seen that the phase is approximately π (Fig 8.3(b)). Thus, the use of the indicator as the slope of a slope is rather limited (cf Eqn (8.4)). The amplitude responses of the combination of the acceleration indicator with the exponential moving average of length 3, and the combination of the acceleration indicator with the exponential moving average of length 6 are also plotted in Fig 8.3(a). Furthermore, the phase responses of the combination of the acceleration indicator with the exponential moving average of length 3, and the combination of the acceleration indicator with the exponential moving average of length 6 are also plotted in Fig 8.3(b). Fig 8.3(b) shows that when the indicator is applied to the smoothed data, the less than ideal phase lag is made even worse.

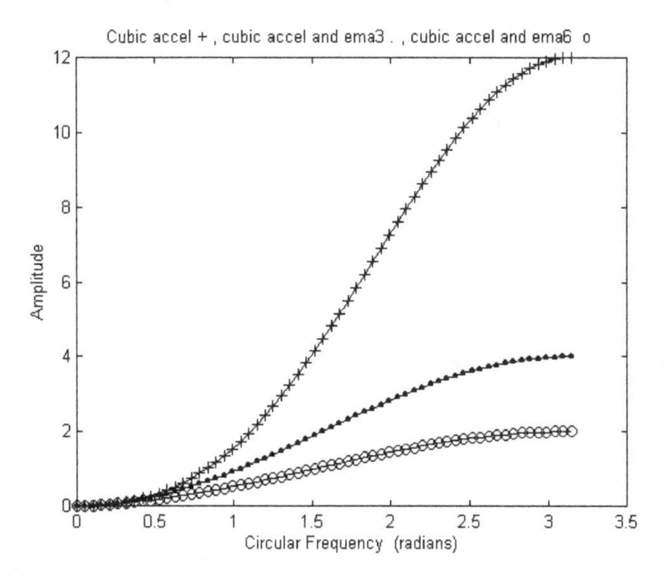

Fig 8.3(a) The amplitude responses of the cubic acceleration indicator (plotted as +), the combination of the cubic acceleration indicator with the exponential moving average of length 3 (plotted as .), and the combination of the cubic acceleration indicator with the exponential moving average of length 6 (plotted as o) are plotted versus ω, the circular frequency.

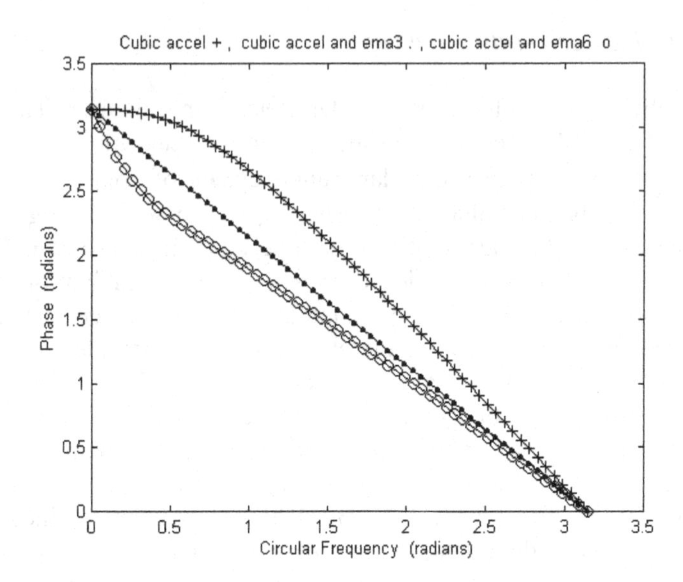

Fig 8.3(b) The phase responses of the cubic acceleration indicator (plotted as +), the combination of the cubic acceleration indicator with the exponential moving average of length 3 (plotted as .), and the combination of the cubic acceleration indicator with the exponential moving average of length 6 (plotted as o) are plotted versus ω, the circular frequency.

In the following sections, we will take a look at higher order polynomials, and see whether the phase lag can be improved.

8.4 Quartic Indicators

8.4.1 *The Filters*

8.4.1.1 *Quartic Velocity Indicator*

The quartic velocity and acceleration indicators will be introduced here. Their derivations are given in Appendix 4. Basically, five adjacent market price data points are fitted to a quartic function. The slope and the slope of the slope of the quartic function at the most recent data point are then calculated using Calculus. They would represent the velocity and acceleration of the market price.

The quartic velocity indicator is defined as (25/12, -4, 3, -4/3, 1/4). The output response, y, after the input price data, x, is filtered by the quartic velocity indicator is

$$y(n) = \frac{25}{12}x(n) - 4x(n-1) + 3x(n-2) - \frac{4}{3}x(n-3) + \frac{1}{4}x(n-4) \quad (8.15)$$

Therefore, the current velocity is given by

$$y(0) = \frac{25}{12}x(0) - 4x(-1) + 3x(-2) - \frac{4}{3}x(-3) + \frac{1}{4}x(-4) \quad (8.16)$$

where x(0) is the closing price or the smoothed closing price of the current bar,

x(-1) is the closing price or the smoothed closing price of one bar ago,

x(-2) is the closing price or the smoothed closing price of two bars ago,

x(-3) is the closing price or the smoothed closing price of three bars ago,

and x(-4) is the closing price or the smoothed closing price of four bars ago.

Fig 8.4(a) shows a market price data simulated as a sine wave. The slope of the sine wave, as calculated using the first derivative in Calculus, is plotted in Fig 8.4(b). The sine wave filtered by the quartic velocity indicator is also plotted, and it agrees quite well with the slope.

In the EasyLanguage code of Omega Research's TradeStation2000i, the program for calculating the quartic velocity indicator can be written as follows: -

```
Input:S(3);
Plot1(25*AMAFUNC2(c,S)/12-
4*AMAFUNC2(c[1],S)+3*AMAFUNC2(c[2],S)-
4*AMAFUNC2(c[3],S)/3+AMAFUNC2(c[4],S)/4,"Plot1");
Plot2(0,"Plot2");
```

c represents the closing price of the current bar. c[1] represents the closing price of one bar ago, c[2] represents the closing price of two bars ago, c[3] represents the closing price of three bars ago and c[4] represents the closing price of four bars ago. AMAFUNC2 is the adaptive moving average function written by Jurik Research. The first input parameter of AMAFUNC2 signifies the closing price series to be smoothed, while the second input parameter, S, indicates the smoothness factor. The larger the smoothness factor, the more smoothed the smoothed data will be. The first line of the program shows that the user can input the smoothness factor, S. Otherwise, it is 3 by default. Two plots are drawn. The first one calculates the quartic velocity of closing prices smoothed by a factor of S. The second one plots a horizontal straight line where the velocity is zero. This straight line helps to identify when the calculated velocity is positive or negative.

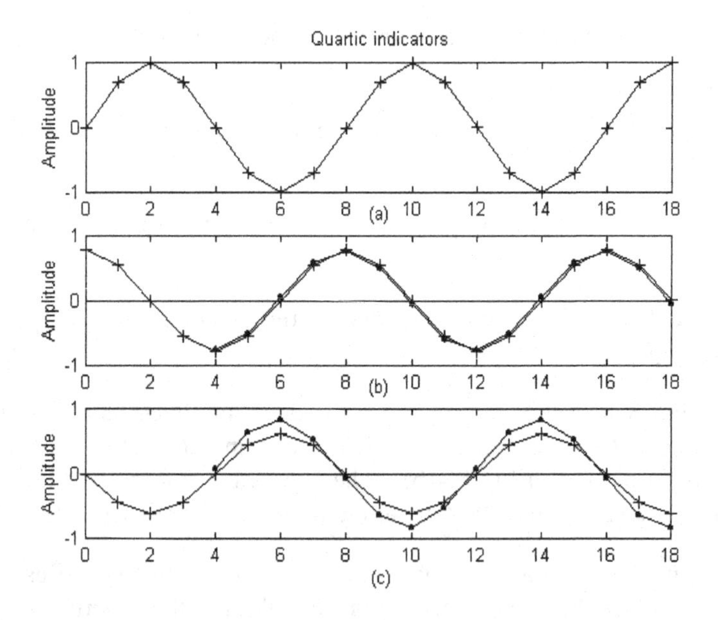

Fig 8.4(a) Market price data simulated as a sine wave of circular frequency of $\pi/4$ radian (b) the sine wave filtered by the quartic velocity indicator (marked as .), is compared with the slope of the sine wave (marked as +) (c) the sine wave filtered by the quartic acceleration indicator (marked as .), is compared with the slope of the slope of the sine wave (marked as +). The first 4 points of the indicator response are not plotted in (b) and (c) as it takes five points to perform the calculations.

AMAFUNC2 can be substituted by other smoothing function. For example, it can be substituted by XAVERAGE, which is a build-in exponential moving average function written by TradeStation2000i. The program plotting the quartic velocity indicators calculated on closing price data smoothed by exponential moving average is listed as follows: -

```
Input:L(6);
Plot1(25*XAVERAGE(c,L)/12-
4*XAVERAGE(c[1],L)+3*XAVERAGE(c[2],L)-
4*XAVERAGE(c[3],L)/3+XAVERAGE(c[4],L)/4,"Plot1");
Plot2(0,"Plot2");
```

The first input parameter of XAVERAGE signifies the closing price series to be smoothed, while the second input parameter, L, indicates the length of the window [Mak 2003]. The larger the length, the more smoothed the smoothed data will be. The first line of the program shows that the user can input the length, L. Otherwise, it is 6 by default. Two plots are drawn. The first one calculates the velocity of closing prices smoothed by a length of L. The second one plots a horizontal straight line where the velocity is zero.

8.4.1.2 *Quartic Acceleration Indicator*

The quartic accelerator indicator is defined as (35/12, -26/3, 19/2, -14/3, 11/12). The output response, y, after the input price data, x, is filtered by the quartic accelerator indicator is

$$y(n)=\frac{35}{12}x(n)-\frac{26}{3}x(n-1)+\frac{19}{2}x(n-2)-\frac{14}{3}x(n-3)+\frac{11}{12}x(n-4) \tag{8.17}$$

Therefore, the current acceleration is given by

$$y(0)=\frac{35}{12}x(0)-\frac{26}{3}x(-1)+\frac{19}{2}x(-2)-\frac{14}{3}x(-3)+\frac{11}{12}x(-4) \tag{8.18}$$

Fig 8.4(c) plots the slope of the slope of the sine wave as calculated from the second derivative in Calculus. The sine wave filtered by the quartic acceleration indicator is also plotted and it agrees reasonably well with the slope of the slope. Even though the filtered wave has a slightly larger amplitude, there is practically no phase lag compared to the slope of the slope.

In the EasyLanguage code of Omega Research's TradeStation2000i, the program for calculating the quartic acceleration indicator can be written as follows: -

```
Input:S(3);
Plot1(35*AMAFUNC2(c,S)/12-
26*AMAFUNC2(c[1],S)/3+19*AMAFUNC2(c[2],S)/2-
14*AMAFUNC2(c[3],S)/3+11*AMAFUNC2(c[4],S)/12,"Plot1");
Plot2(0,"Plot2");
```

The first line of the program shows that the user can input the smoothness factor, S. Otherwise, it is 3 by default. Two plots are drawn. The first one calculates the quartic acceleration of closing prices smoothed by a factor S. The second one plots a horizontal straight line where the acceleration is zero. AMAFUNC2 can be substituted by other smoothing function. For example, it can be substituted by XAVERAGE. The program plotting the quartic acceleration indicators calculated on closing price data smoothed by exponential moving average is listed as follows: -

```
Input:L(6);
Plot1(35*XAVERAGE(c,L)/12-
26*XAVERAGE(c[1],L)/3+19*XAVERAGE(c[2],L)/2-
14*XAVERAGE(c[3],L)/3+11*XAVERAGE(c[4],L)/12,"Plot1");
Plot2(0,"Plot2");
```

The first line of the program shows that the user can input the length, L. Otherwise, it is 6 by default. Two plots are drawn. The first one calculates the acceleration of closing prices smoothed by a length of L. The second one plots a horizontal straight line where the acceleration is zero.

An example using the quartic velocity and acceleration indicators are shown in Fig 8.5. The figure shows a weekly chart of the S&P 500 Index. The price data was smoothed using the adaptive moving average of Jurik research with smoothness 3 (shown as a line in the top figure). Quartic velocity indicator was applied to the moving average and plotted in the middle of the figure. Quartic acceleration indicator was also applied to the moving average and plotted in the bottom of the figure. The figure shows the S&P 500 Index between May 2002 and January 2004. The S&P 500 Index achieved an all time high of 1530 in Sept 2000 and fell to a triple bottom of approximately 775 in 2002 and 2003, as shown in Fig 8.5. The magnitude of the velocities of the triple bottom, as plotted in the middle plot of Fig 8.5, are getting smaller and smaller, showing a divergence with the price. This implies that the market will rise after [Mak 2003]. As it happened, the market did rise. A good buying opportunity would occur when the velocity is zero right after the the third bottom. Divergence will be discussed in more detail in Chapter 10.

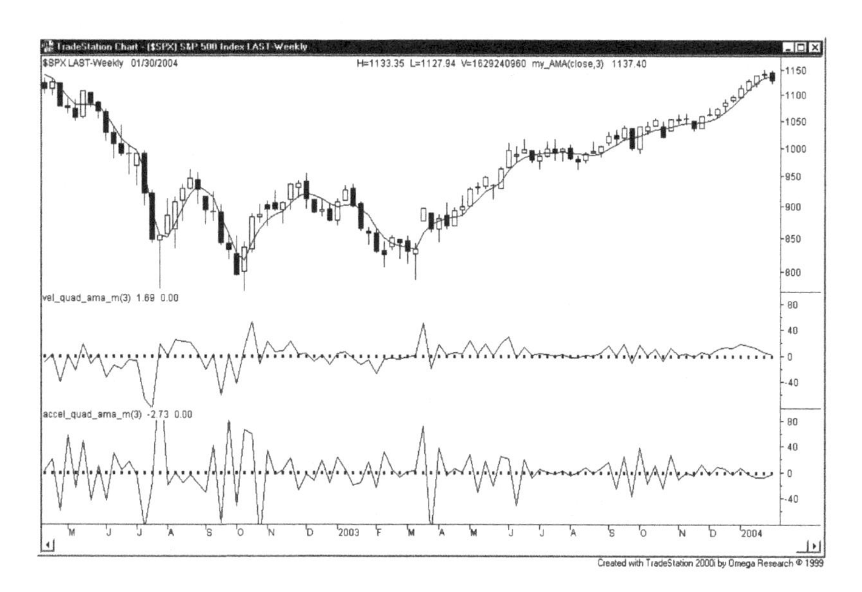

Fig 8.5 A weekly chart of the S&P 500 Index. The price data was smoothed using the adaptive moving average with smoothness 3 (shown as a line in the top figure). Quartic velocity and quartic acceleration indicators are plotted in the middle and bottom figures respectively. *Chart produced with Omega Research TradeStation2000i.*

8.4.2 *Filtering Smoothed Data*

8.4.2.1 *Quartic Velocity Indicator*

The frequency response or the Discrete Time Fourier Transform (DTFT) of the quartic velocity indicator is given by

$H(\omega) =$

$$25/12 - 4\exp(-i\omega) + 3\exp(-2i\omega) - (4/3)\exp(-3i\omega) + (1/4)\exp(-4i\omega) \quad (8.19)$$

The amplitude and phase of $H(\omega)$ is plotted respectively in Fig 8.6(a) and (b). The ideal phase lead for a velocity indicator should be $\pi/2$ for all frequencies. As can be seen from Fig 8.6(b), the quartic velocity indicator maintains a $\pi/2$ phase lead for frequency up to approximately 1 radian, rises slowly and then drops off below $\pi/2$ at about 2 radians. Thus, the indicator provides a reasonable velocity indicator for trading purpose.

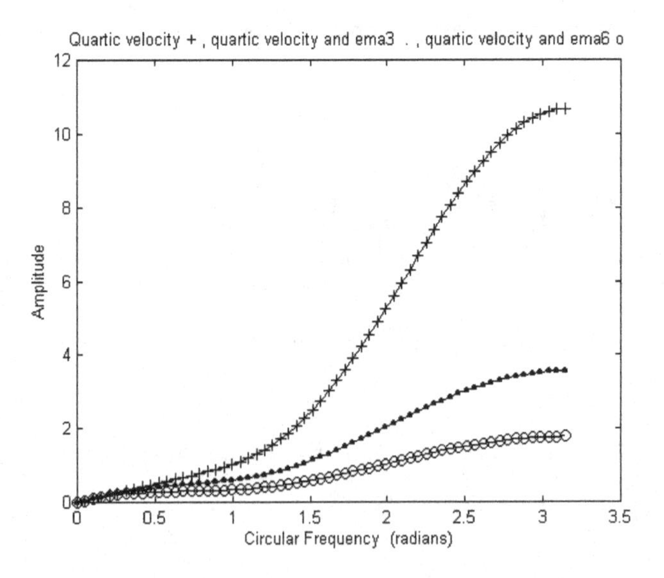

Fig 8.6(a) The amplitude responses of the quartic velocity indicator (plotted as +), the combination of the quartic velocity indicator with the exponential moving average of length 3 (plotted as .), and the combination of the quartic velocity indicator with the exponential moving average of length 6 (plotted as o) are plotted versus ω, the circular frequency.

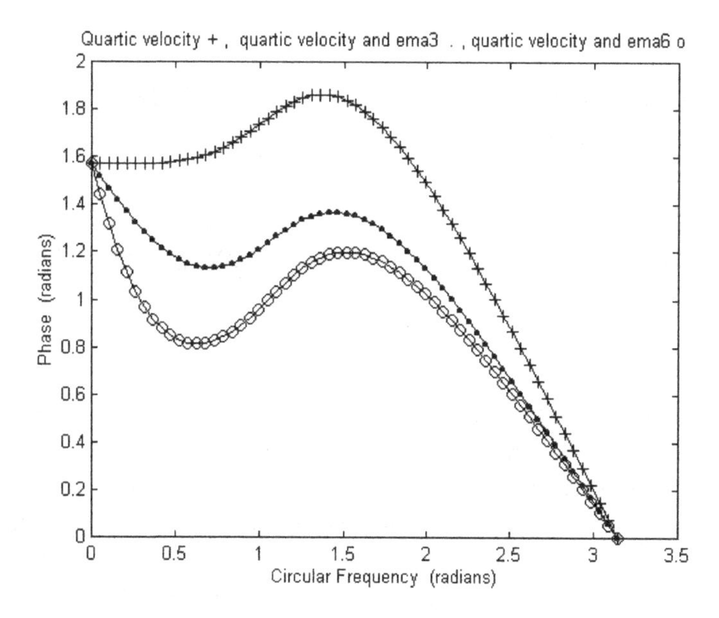

Fig 8.6(b) The phase responses of the quartic velocity indicator (plotted as +), the combination of the quartic velocity indicator with the exponential moving average of length 3 (plotted as .), and the combination of the quartic velocity indicator with the exponential moving average of length 6 (plotted as o) are plotted versus ω, the circular frequency.

The amplitude responses of the combination of the velocity indicator with the exponential moving average of length 3, and the combination of the velocity indicator with the exponential moving average of length 6 are also plotted in Fig 8.6(a). Furthermore, the phase responses of the combination of the velocity indicator with the exponential moving average of length 3, and the combination of the velocity indicator with the exponential moving average of length 6 are also plotted in Fig 8.6(b). Fig 8.6(b) shows that the combined phase lag is somewhat less than $\pi/2$, but it is not so far off as to make it unsuitable for measuring slope. The phase lag would improve if an adaptive moving average were used instead of the exponential moving average.

8.4.2.2 *Quartic Acceleration Indicator*

The frequency response or the Discrete Time Fourier Transform (DTFT) of the quartic acceleration indicator is given by

$$H(\omega) =$$

$$35/12 - (26/3)\exp(-i\omega) + (19/2)\exp(-2i\omega) - (14/3)\exp(-3i\omega) + (11/12)\exp(-4i\omega) \tag{8.20}$$

The amplitude and phase of H(ω) is plotted respectively in Fig 8.7(a) and (b). The ideal phase lead for an acceleration indicator should be π for all frequencies. As can be seen from Fig 8.7(b), the quartic acceleration indicator maintains a π phase lead for frequency up to approximately 1.2 radian before dropping off. Thus, the indicator provides a reasonable acceleration indicator for trading purpose.

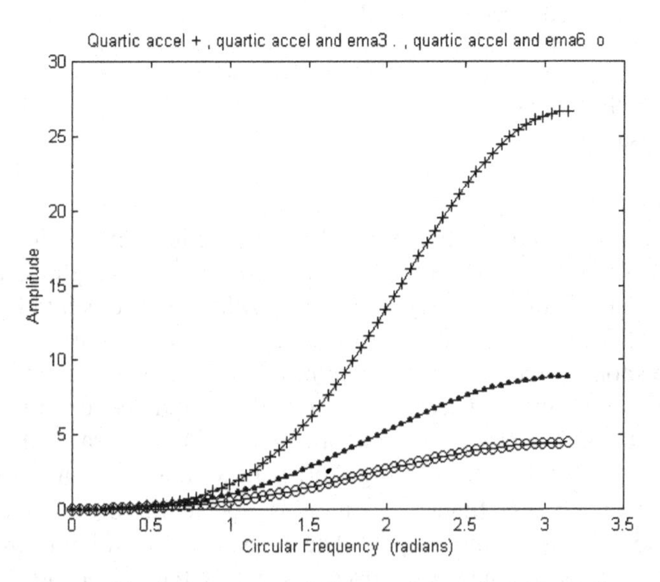

Fig 8.7(a) The amplitude responses of the quartic acceleration indicator (plotted as +), the combination of the quartic acceleration indicator with the exponential moving average of length 3 (plotted as .), and the combination of the quartic acceleration indicator with the exponential moving average of length 6 (plotted as o) are plotted versus ω, the circular frequency.

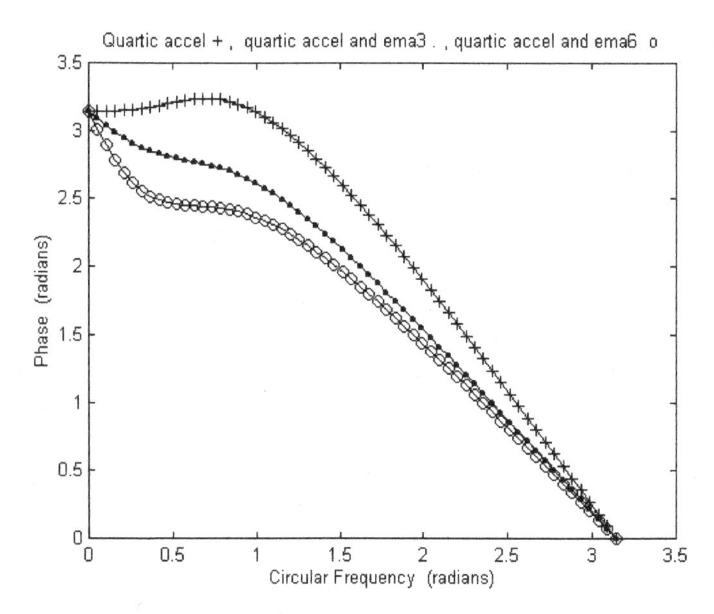

Fig 8.7(b) The phase responses of the quartic acceleration indicator (plotted as +), the combination of the quartic acceleration indicator with the exponential moving average of length 3 (plotted as .), and the combination of the quartic acceleration indicator with the exponential moving average of length 6 (plotted as o) are plotted versus ω, the circular frequency.

The amplitude responses of the combination of the acceleration indicator with the exponential moving average of length 3, and the combination of the acceleration indicator with the exponential moving average of length 6 are also plotted in Fig 8.7(a). Furthermore, the phase responses of the combination of the acceleration indicator with the exponential moving average of length 3, and the combination of the acceleration indicator with the exponential moving average of length 6 are also plotted in Fig 8.7(b). Fig 8.7(b) shows that the combined phase lag is somewhat less than π, but it is not so far off as to make it unsuitable for measuring slope of the slope. The phase lag would improve if an adaptive moving average were used instead of the exponential moving average.

8.5 Quintic Indicators

8.5.1 *The Filters*

8.5.1.1 *Quintic Velocity Indicator*

The quintic velocity and acceleration indicators will be introduced here. Their derivations are given in Appendix 4. Basically, six adjacent market price data points are fitted to a quintic function. The slope and the slope of the slope of the quintic function at the most recent data point are then calculated using Calculus. They would represent the velocity and acceleration of the market price.

The quintic velocity indicator is defined as (137/60 -5 5 -10/3 5/4 -1/5). The output response, y, after the input price data, x, is filtered by the quinic velocity indicator is

$$y(n) = \frac{137}{60}x(n) - 5x(n-1) + 5x(n-2) - \frac{10}{3}x(n-3) + \frac{5}{4}x(n-4) - \frac{1}{5}x(n-5) \tag{8.21}$$

Therefore, the current velocity is given by

$$y(0) = \frac{137}{60}x(0) - 5x(-1) + 5x(-2) - \frac{10}{3}x(-3) + \frac{5}{4}x(-4) - \frac{1}{5}x(-5) \tag{8.22}$$

where x(0) is the closing price or the smoothed closing price of the current bar,

x(-1) is the closing price or the smoothed closing price of one bar ago,

x(-2) is the closing price or the smoothed closing price of two bars ago, x(-3) is the closing price or the smoothed closing price of three bars ago,

x(-4) is the closing price or the smoothed closing price of four bars ago,

and x(-5) is the closing price or the smoothed closing price of five bars ago.

Fig 8.8(a) shows a market price data simulated as a sine wave. The slope of the sine wave, as calculated from Calculus, is plotted in Fig 8.8(b). The sine wave filtered by the quintic velocity indicator is also plotted, and it agrees quite well with the slope.

8.5.1.2 *Quintic Acceleration Indicator*

The quintic accelerator indicator is defined as (15/4 -77/6 107/6 -13 61/12 -5/6). The output response, y, after the input price data, x, is filtered by the quintic accelerator indicator is

$$y(n) = \frac{15}{4}x(n) - \frac{77}{6}x(n-1) + \frac{107}{6}x(n-2) - 13x(n-3) + \frac{61}{12}x(n-4) - \frac{5}{6}x(n-5) \qquad (8.23)$$

Therefore, the current acceleration is given by

$$y(0) = \frac{15}{4}x(0) - \frac{77}{6}x(-1) + \frac{107}{6}x(-2) - 13x(-3) + \frac{61}{12}x(-4) - \frac{5}{6}x(-5) \qquad (8.24)$$

Fig 8.8(c) plots the slope of the slope of the sine wave as calculated from Calculus. The sine wave filtered by the quintic acceleration indicator is also plotted and it agrees reasonably well with the slope of the slope. Even though the filtered wave has a slightly larger amplitude, there is practically no phase lag compared to the slope of the slope.

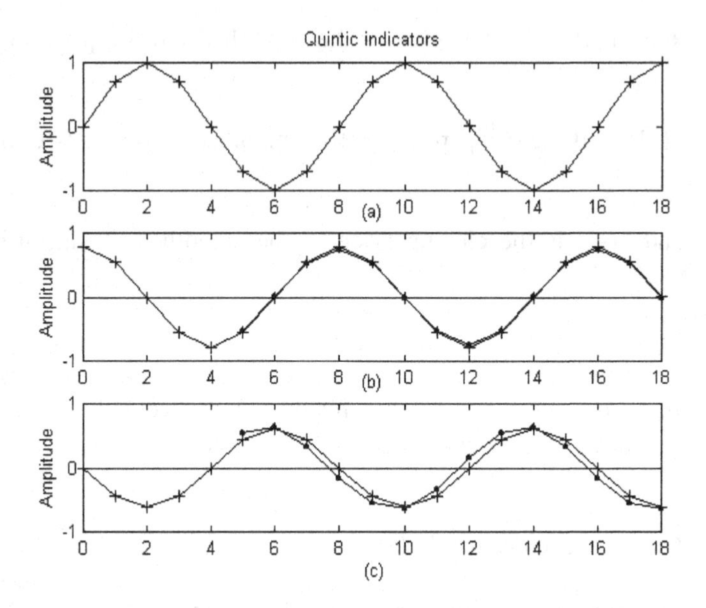

Fig 8.8(a) Market price data simulated as a sine wave of circular frequency of $\pi/4$ radian (b) the sine wave filtered by the quintic velocity indicator (marked as .), is compared with the slope of the sine wave (marked as +) (c) the sine wave filtered by the quintic acceleration indicator (marked as .), is compared with the slope of the slope of the sine wave (marked as +). The first 5 points of the indicator response are not plotted in (b) and (c) as it takes six points to perform the calculations.

8.5.2 *Filtering Smoothed Data*

8.5.2.1 *Quintic Velocity Indicator*

The frequency response or the Discrete Time Fourier Transform (DTFT) of the quintic velocity indicator is given by

$$H(\omega) = 137/60 - 5\exp(-i\omega) + 5\exp(-2i\omega) - (10/3)\exp(-3i\omega) + (5/4)\exp(-4i\omega) - (1/5)\exp(-5i\omega) \quad (8.25)$$

The amplitude and phase of $H(\omega)$ is plotted respectively in Fig 8.9(a) and (b). The ideal phase lead for a velocity indicator should be $\pi/2$ for all frequencies. As can be seen from Fig 8.9(b), the quintic velocity indicator maintains a $\pi/2$ phase lead for frequency up to approximately 1 radian, rises to about 2.2 and then drops off below $\pi/2$ at about 2 radians. Thus, the indicator provides a reasonable velocity indicator for trading purpose.

The amplitude responses of the combination of the velocity indicator with the exponential moving average of length 3, and the combination of the velocity indicator with the exponential moving average of length 6 are also plotted in Fig 8.9(a). Furthermore, the phase responses of the combination of the velocity indicator with the exponential moving average of length 3, and the combination of the velocity indicator with the exponential moving average of length 6 are also plotted in Fig 8.9(b). Fig 8.9(b) shows that the combined phase lag is somewhat close to $\pi/2$, so it is a reasonable indicator for measuring slope. The phase lag would improve if an adaptive moving average were used instead of the exponential moving average.

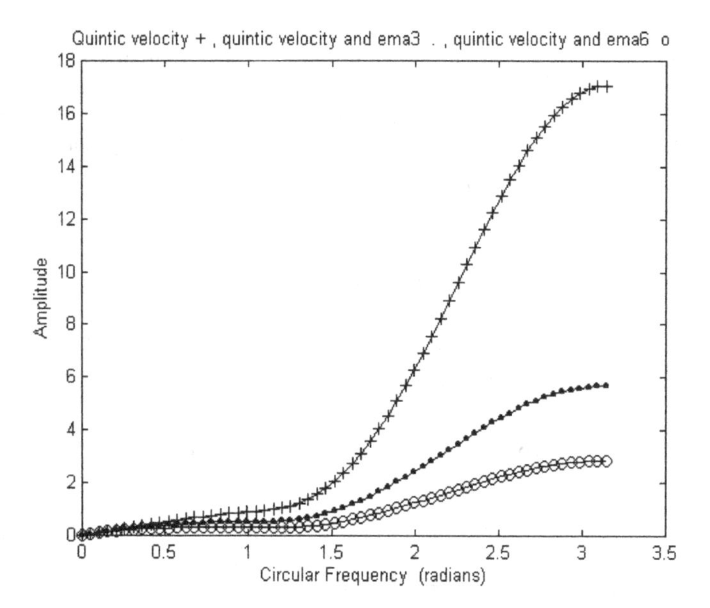

Fig 8.9(a) The amplitude responses of the quintic velocity indicator (plotted as +), the combination of the quintic velocity indicator with the exponential moving average of length 3 (plotted as .), and the combination of the quintic velocity indicator with the exponential moving average of length 6 (plotted as o) are plotted versus ω, the circular frequency.

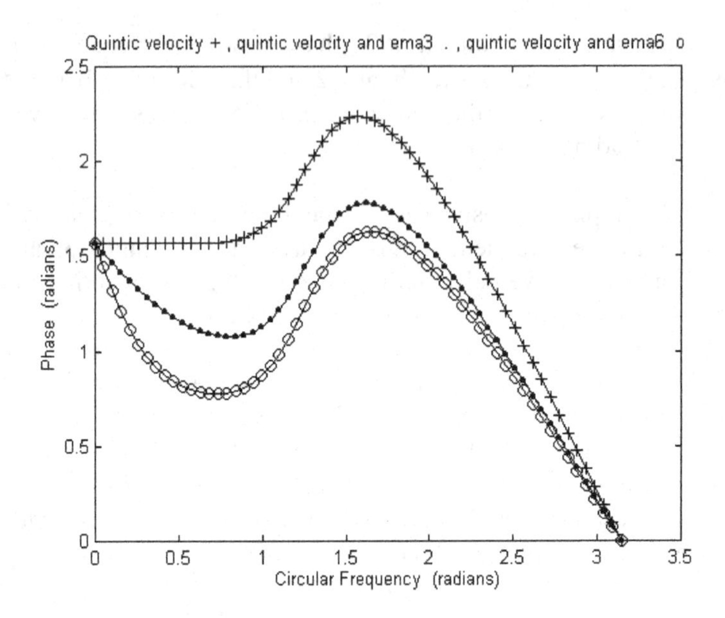

Fig 8.9(b) The phase responses of the quintic velocity indicator (plotted as +), the combination of the quintic velocity indicator with the exponential moving average of length 3 (plotted as .), and the combination of the quintic velocity indicator with the exponential moving average of length 6 (plotted as o) are plotted versus ω, the circular frequency.

8.5.2.2 *Quintic Acceleration Indicator*

The frequency response or the Discrete Time Fourier Transform (DTFT) of the quintic acceleration indicator is given by

$$H(\omega) = 15/4 - (77/6)\exp(-i\omega) - (107/6)\exp(-2i\omega) - 13\exp(-3i\omega)$$

$$+ (61/12)\exp(-4i\omega) - (5/6)\exp(-5i\omega) \qquad (8.26)$$

The amplitude and phase of H(ω) is plotted respectively in Fig 8.10(a) and (b). The ideal phase lead for an acceleration indicator should be π for all frequencies. As can be seen from Fig 8.10(b), the quintic acceleration indicator maintains a π phase lead for frequency up to approximately 0.5 radian, rises to approximately 3.6, and then drops off below π at about 1.5 radians. Thus, the indicator provides a reasonable acceleration indicator for trading purpose.

The amplitude responses of the combination of the acceleration indicator with the exponential moving average of length 3, and the combination of the acceleration indicator with the exponential moving average of length 6 are also plotted in Fig 8.10(a). Furthermore, the phase responses of the combination of the acceleration indicator with the exponential moving average of length 3, and the combination of the acceleration indicator with the exponential moving average of length 6 are also plotted in Fig 8.10(b). Fig 8.10(b) shows that the combined phase lag is slightly less than π, but it is reasonably suitable for measuring the slope of the slope. The phase lag would improve if an adaptive moving average were used instead of the exponential moving average.

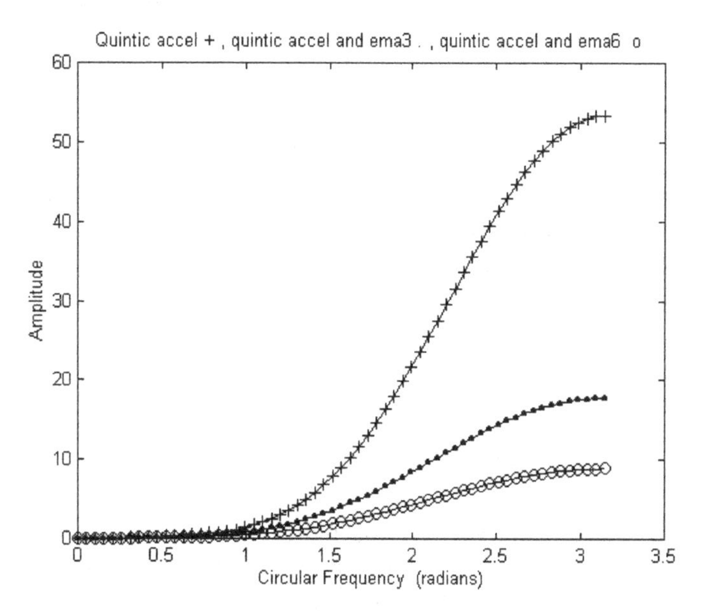

Fig 8.10(a) The amplitude responses of the quintic acceleration indicator (plotted as +), the combination of the quintic acceleration indicator with the exponential moving average of length 3 (plotted as .), and the combination of the quintic acceleration indicator with the exponential moving average of length 6 (plotted as o) are plotted versus ω, the circular frequency.

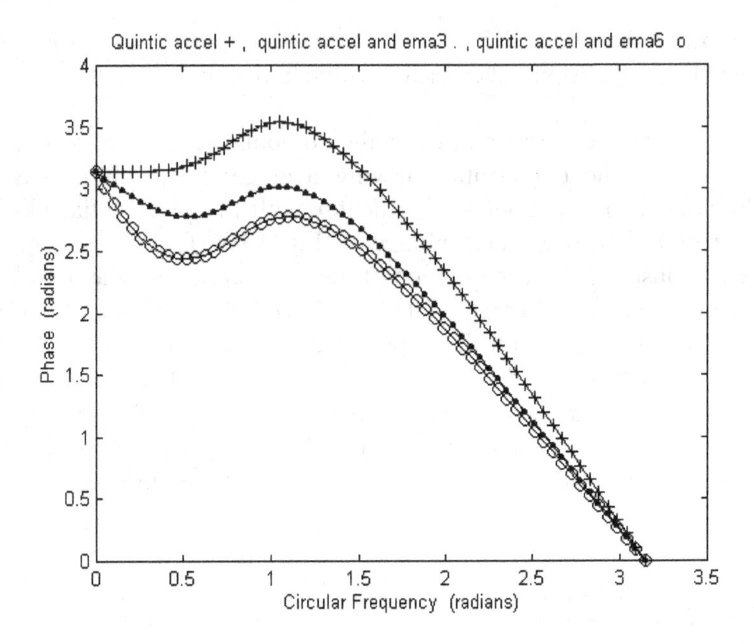

Fig 8.10(b) The phase responses of the quintic acceleration indicator (plotted as +), the combination of the quintic acceleration indicator with the exponential moving average of length 3 (plotted as .), and the combination of the quintic acceleration indicator with the exponential moving average of length 6 (plotted as o) are plotted versus ω, the circular frequency.

8.6 Sextic Indicators

8.6.1 *The Filters*

8.6.1.1 *Sextic Velocity Indicator*

The sextic velocity and acceleration indicators will be introduced here. Their derivations are given in Appendix 4. Basically, seven adjacent market price data points are fitted to a sextic function. The slope and the slope of the slope of the sextic function at the most recent data point are then calculated using Calculus. They would represent the velocity and acceleration of the market price.

The sextic velocity indicator is defined as (49/20 -6 15/2 -20/3 15/4 -6/5 1/6). The output response, y, after the input price data, x, is filtered by the sextic velocity indicator is

$$y(n) = \frac{49}{20}x(n) - 6x(n-1) + \frac{15}{2}x(n-2) - \frac{20}{3}x(n-3) + \frac{15}{4}x(n-4) - \frac{6}{5}x(n-5) + \frac{1}{6}x(n-6) \qquad (8.27)$$

Therefore, the current velocity is given by

$$y(0) = \frac{49}{20}x(0) - 6x(-1) + \frac{15}{2}x(-2) - \frac{20}{3}x(-3) + \frac{15}{4}x(-4) - \frac{6}{5}x(-5) + \frac{1}{6}x(-6) \qquad (8.28)$$

where x(0) is the closing price or the smoothed closing price of the current bar,

x(-1) is the closing price or the smoothed closing price of one bar ago,

x(-2) is the closing price or the smoothed closing price of two bars ago,

x(-3) is the closing price or the smoothed closing price of three bars ago,

x(-4) is the closing price or the smoothed closing price of four bars ago,

x(-5) is the closing price or the smoothed closing price of five bars ago,

and x(-6) is the closing price or the smoothed closing price of six bars ago.

Fig 8.11(a) shows a market price data simulated as a sine wave. The slope of the sine wave, as calculated from Calculus, is plotted in Fig 8.11(b). The sine wave filtered by the sextic velocity indicator is also plotted, and it agrees quite well with the slope.

8.6.1.2 *Sextic Acceleration Indicator*

The sextic acceleration indicator is defined as (4.5112 -17.4 29.25 -28.2222 16.5 -5.4 0.7612). The output response, y, after the input price data, x, is filtered by the sextic accelerator indicator is

$$y(n) = 4.5112x(n) - 17.4x(n-1) + 29.25x(n-2) - 28.2222x(n-3)$$

$$+16.5x(n-4) - 5.4x(n-5) + 0.7612x(n-6) \qquad (8.29)$$

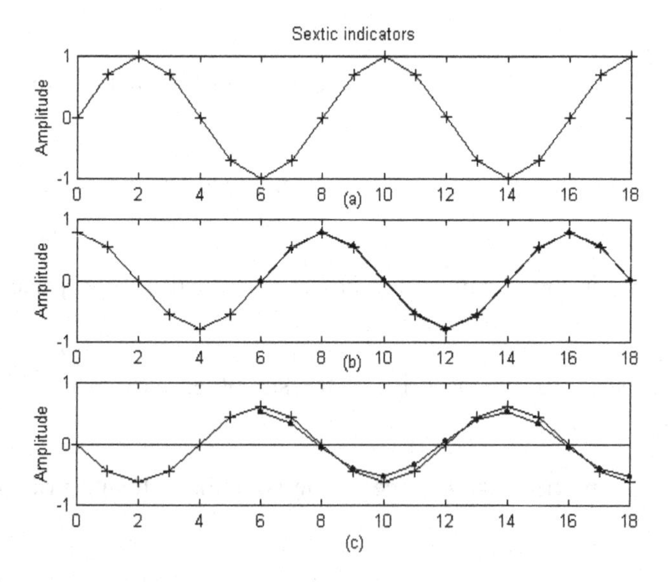

Fig 8.11(a) Market price data simulated as a sine wave of circular frequency of π/4 radian (b) the sine wave filtered by the sextic velocity indicator (marked as .), is compared with the slope of the sine wave (marked as +) (c) the sine wave filtered by the sextic acceleration indicator (marked as .), is compared with the slope of the slope of the sine wave (marked as +). The first 6 points of the indicator response are not plotted in (b) and (c) as it takes seven points to perform the calculations.

Therefore, the current acceleration is given by

$$y(0) = 4.5112x(0) - 17.4x(-1) + 29.25x(-2) - 28.2222x(-3)$$

$$+16.5x(-4) - 5.4x(-5) + 0.7612x(-6) \qquad (8.30)$$

Fig 8.11(c) plots the slope of the slope of the sine wave as calculated from Calculus. The sine wave filtered by the sextic acceleration indicator is also plotted and it agrees reasonably well with the slope of the slope. Even though the filtered wave has a slightly smaller amplitude, there is practically no phase lag compared to the slope of the slope.

8.6.2 *Filtering Smoothed Data*

8.6.2.1 *Sextic Velocity Indicator*

The frequency response or the Discrete Time Fourier Transform (DTFT) of the sextic velocity indicator is given by

$$H(\omega) = 49/20 - 6\exp(-i\omega) + (15/2)\exp(-2i\omega) - (20/3)\exp(-3i\omega)$$

$$+(15/4)\exp(-4i\omega) - (6/5)\exp(-5i\omega) + (1/6)\exp(-6i\omega) \qquad (8.31)$$

The amplitude and phase of $H(\omega)$ is plotted respectively in Fig 8.12(a) and (b). The ideal phase lead for a velocity indicator should be $\pi/2$ for all frequencies. As can be seen from Fig 8.12(b), the sextic velocity indicator maintains a $\pi/2$ phase lead for frequency up to approximately 1.2 radian, rises to about 2.8 and then drops off below $\pi/2$ at about 2.3 radians. Thus, while the indicator is a good velocity indicator for low frequencies, it is not a good velocity indicator for high frequencies.

The amplitude responses of the combination of the velocity indicator with the exponential moving average of length 3, and the combination of the velocity indicator with the exponential moving average of length 6 are also plotted in Fig 8.12(a). Furthermore, the

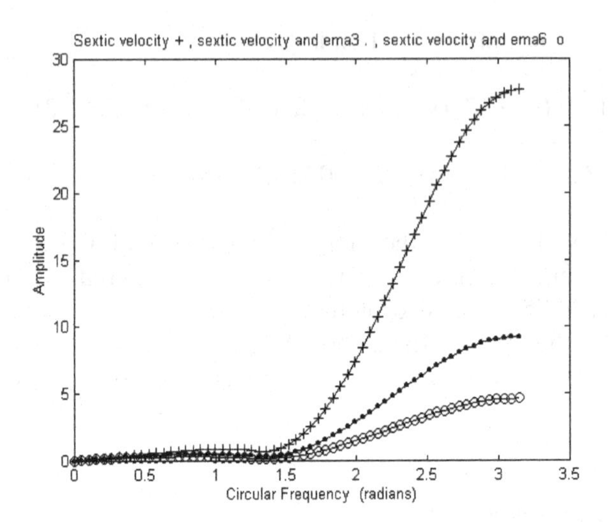

Fig 8.12(a) The amplitude responses of the sextic velocity indicator (plotted as +), the combination of the sextic velocity indicator with the exponential moving average of length 3 (plotted as .), and the combination of the sextic velocity indicator with the exponential moving average of length 6 (plotted as o) are plotted versus ω, the circular frequency.

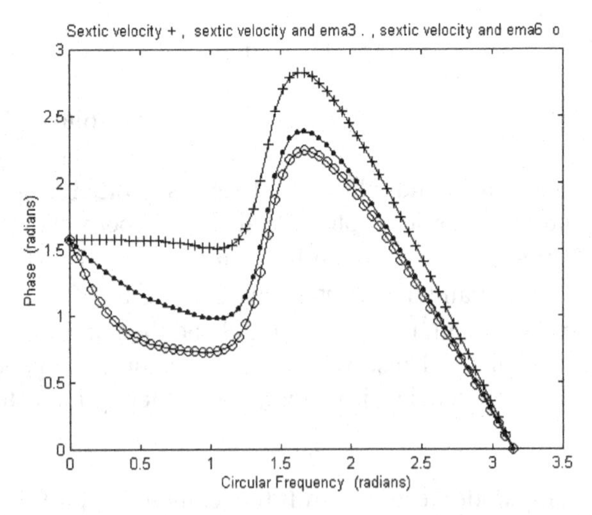

Fig 8.12(b) The phase responses of the sextic velocity indicator (plotted as +), the combination of the sextic velocity indicator with the exponential moving average of length 3 (plotted as .), and the combination of the sextic velocity indicator with the exponential moving average of length 6 (plotted as o) are plotted versus ω, the circular frequency.

phase responses of the combination of the velocity indicator with the exponential moving average of length 3, and the combination of the velocity indicator with the exponential moving average of length 6 are also plotted in Fig 8.12(b). Fig 8.12(b) shows that the combined phase lag fluctuates about $\pi/2$, so it is not a particular good indicator for measuring slope

8.6.2.2 *Sextic Acceleration Indicator*

The frequency response or the Discrete Time Fourier Transform (DTFT) of the sextic acceleration indicator is given by

$$H(\omega) = 4.5112 - 17.4\exp(-i\omega) - 29.25\exp(-2i\omega) - 28.2222\exp(-3i\omega)$$

$$+ 16.5\exp(-4i\omega) - 5.4\exp(-5i\omega) + 0.7612\exp(-6i\omega) \qquad (8.32)$$

The amplitude and phase of $H(\omega)$ is plotted respectively in Fig 8.13(a) and (b). The ideal phase lead for an acceleration indicator should be π for all frequencies. As can be seen from Fig 8.13(b), the sextic acceleration indicator maintains a π phase lead for frequency up to approximately 1 radian, rises to approximately 4.2, and then drops off below π at about 2 radians. Thus, the indicator provides a reasonable acceleration indicator for trading purpose.

The amplitude responses of the combination of the acceleration indicator with the exponential moving average of length 3, and the combination of the acceleration indicator with the exponential moving average of length 6 are also plotted in Fig 8.13(a). Furthermore, the phase responses of the combination of the acceleration indicator with the exponential moving average of length 3, and the combination of the acceleration indicator with the exponential moving average of length 6 are also plotted in Fig 8.13(b). Fig 8.13(b) shows that the combined phase lag averages about π, so it is reasonably suitable for measuring the slope of the slope.

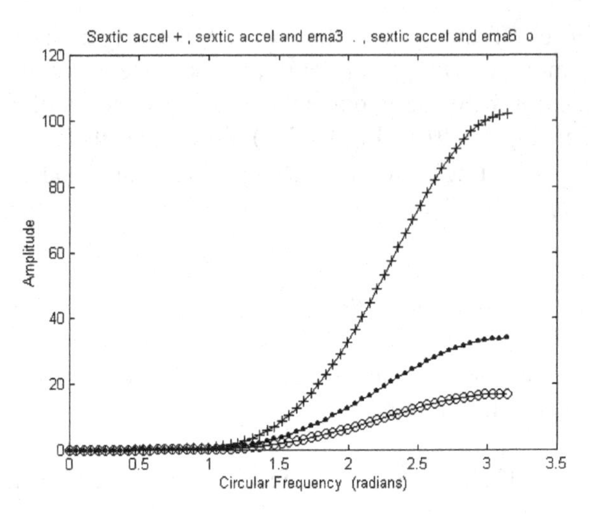

Fig 8.13(a) The amplitude responses of the sextic acceleration indicator (plotted as +), the combination of the sextic acceleration indicator with the exponential moving average of length 3 (plotted as .), and the combination of the sextic acceleration indicator with the exponential moving average of length 6 (plotted as o) are plotted versus ω, the circular frequency.

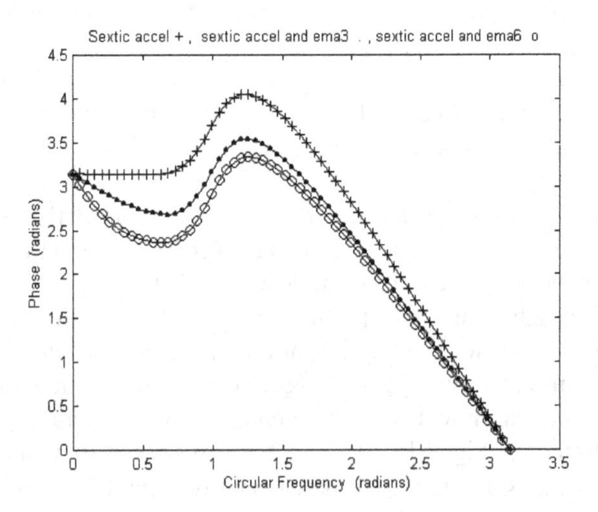

Fig 8.13(b) The phase responses of the quintic acceleration indicator (plotted as +), the combination of the quintic acceleration indicator with the exponential moving average of length 3 (plotted as .), and the combination of the quintic acceleration indicator with the exponential moving average of length 6 (plotted as o) are plotted versus ω, the circular frequency.

8.7 Velocity and Acceleration Indicator Responses on Smoothed Data

Judging from the phase responses of the velocity and acceleration indicators on smoothed data, it appears that the quartic and quintic indicators provide a flatter response with respect to frequency, and thus would be better indicators for trading purposes than the momentum indicator currently used.

Chapter 9

Skipped Convolution

Skipped convolution has been introduced by Mak [2003]. It has the advantage that it can alert traders of a trading opportunity earlier. In this chapter, we will take a look at its frequency response and some of its limitations.

9.1 Frequency Response

9.1.1 *Frequency Response of a Convolution*

We will first take a look at the conventional convolution. The output, y(n), of the convolution of an unit impulse response, h(k), of a Finite Impulse Response (FIR) filter with an input signal x(n) can be written as [Mak 2003]

$$y(n) = \sum_{k=0}^{K} h(k)x(n-k) \tag{9.1}$$

The performance of a filter may be improved by making use of the output values that have already been processed. That is, previous values of y can be used. In that case, the output, y(n) can be written as:

$$y(n) = \sum_{k=0}^{K} h(k)x(n-k) + \sum_{\ell=1}^{L} g(\ell)y(n-\ell) \tag{9.2}$$

If the g coefficients in Eq (9.2) were set to zero, Eq (9.2) would be reduced to Eq (9.1). The first term on the RHS of Eq (9.2) corresponds to the FIR filter, which is sometimes called a non-recursive filter, as it does not make use of previously processed signal. The second term on the RHS of Eqn 9.2 corresponds to the Infinite Impulse Response (IIR) filter, which is sometimes called a recursive filter as it makes use of previously processed values [Broesch 1997].

Fourier Transform of Eq (9.2) would yield a frequency response function $H(\omega)$:

$$H(\omega) = \frac{Y(\omega)}{X(\omega)} = \frac{\sum_{k=0}^{K} h(k) e^{-i\omega k}}{1 - \sum_{\ell=1}^{L} g(\ell) e^{-i\omega \ell}} \tag{9.3}$$

9.1.2 *Frequency Response of a Skipped Convolution*

The output response of a skipped convolution can be written as:

$$y_D(n) = \sum_{k=0}^{K} h(k)x(n - Dk) + \sum_{\ell=1}^{L} g(\ell) y_D(n - D\ell) \tag{9.4}$$

where D is the skip parameter.

Fourier Transform of Eq (9.4) would yield a frequency response function $H_D(\omega)$:

$$H_D(\omega) = \frac{Y_D(\omega)}{X(\omega)} = \frac{\sum_{k=0}^{K} h(k) e^{-iD\omega k}}{1 - \sum_{\ell=1}^{L} g(\ell) e^{-iD\omega \ell}} \tag{9.5}$$

Comparing Eq (9.5) with Eq (9.4), it can be seen that $H_D(\omega)$ is equal to $H(D\omega)$, i.e., $H_D(\omega)$ is equal to $H(\omega)$ compressed by a factor of D.

$$H_D(\omega) = H(D\omega) \quad (9.6)$$

We will take a look at an example in the following section.

9.2 Skipped Exponential Moving Average

The equation for the output response of a skipped EMA is given by

$$y_D(n) = \alpha x(n) + (1-\alpha)y_D(n-D) \quad (9.7)$$

$$\text{where } \alpha = 2/(M+1) \quad (9.8)$$

M is a positive integer chosen by the trader and is often called the length of the EMA,

and D is the skip parameter.

Eq (9.7) makes use of an output response that has already been processed D bars ago. When D = 1, Eq (9.7) reduces to the ordinary exponential moving average.

To calculate the frequency response of the skipped EMA, Eq (9.5) can be used. Alternatively, we can take the z-transform of Eq (9.7) [Broesch 1997, Proakis and Manolakis 1996].

$$Y_D(z) = \alpha X(z) + (1-\alpha)z^{-D}Y_D(z) \quad (9.9)$$

where $z = r\exp(i\omega)$ is a complex number in the complex plane, r being the magnitude of z. $Y_D(z)$ is the transform of the output and $X(z)$ is the transform of the input.

Defining the transfer function as the output of the filter over the input of the filter

$$H_D(z) = Y_D(z)/X(z) \quad (9.10)$$

we get, for EMA

$$H_D(z) = \frac{\alpha}{1-(1-\alpha)z^{-D}} \tag{9.11}$$

The skipped EMA has a single pole in its transfer function. A pole is a zero of the denominator polynomial of the transfer function $H_D(z)$. Restricting z in the complex plane to $\exp(i\omega)$ on the unit circle (i.e. $r = 1$), the frequency response function $H_D(\omega)$ is given by

$$H_D(\omega) = \frac{\alpha}{1-(1-\alpha)\exp(-iD\omega)} \tag{9.12}$$

The magnitude of $H_D(\omega)$ is given by

$$|H_D(\omega)| = \frac{\alpha}{[1-2(1-\alpha)\cos D\omega + (1-\alpha)^2]^{1/2}} \tag{9.13}$$

The phase is given by

$$\phi_D(\omega) = \tan^{-1}\left[\frac{-(1-\alpha)\sin D\omega}{1-(1-\alpha)\cos D\omega}\right] \tag{9.14}$$

The amplitude and phase of $H(\omega)$ of the EMA for $M = 3$ and $M = 6$ are plotted in Fig 9.1 from 0 to 5π. The amplitude and phase of $H_D(\omega)$ of the skipped EMA with $D = 5$ are plotted in Fig 9.2 for $M = 3$ and $M = 6$ from 0 to π. In consistent with Eq (9.6), Fig 9.2 is exactly the same as Fig 9.1 but compressed by a factor of 5. Comparing Fig 9.2 with Fig 9.1, it can be seen that while, in general, the phase lag of the skipped convolution is less than that of the convolution, the skipped convolution can pick up frequencies larger than π from the original input signal, thus making the output signal of the skipped convolution noisier.

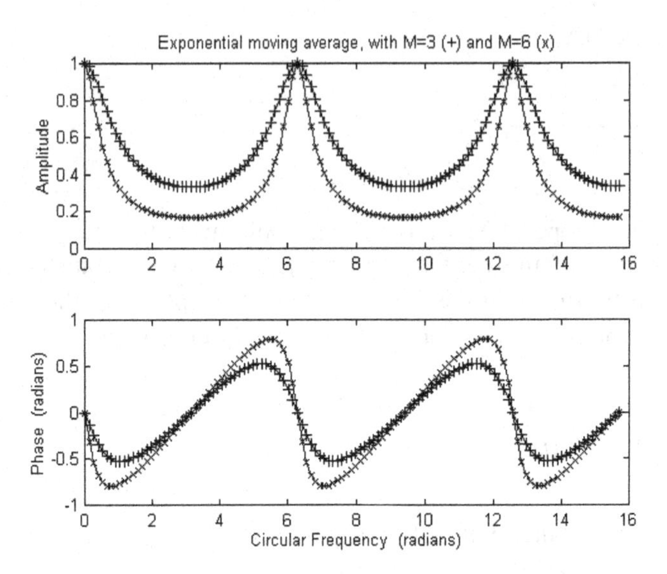

Fig 9.1 The amplitude and phase of the discrete time Fourier Transform, $H(\omega)$, of the exponential moving average for M = 3 (marked as +) and M = 6 (marked as x) are plotted from 0 to 5π.

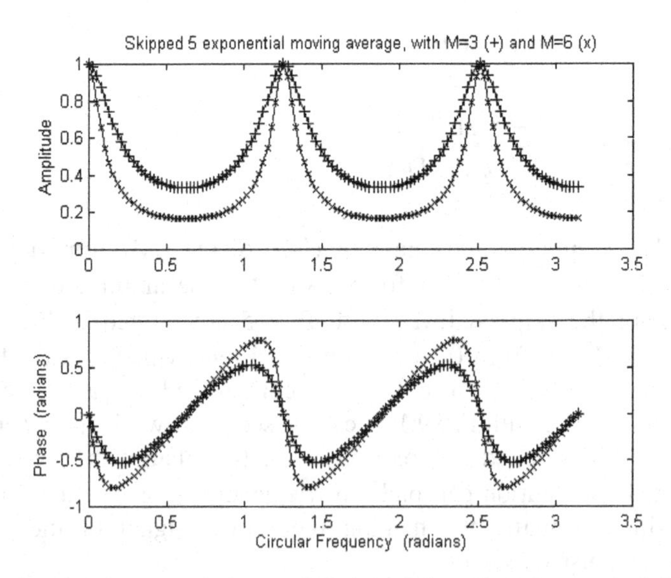

Fig 9.2 The amplitude and phase of $H_D(\omega)$ of the skipped exponential moving average with D = 5 are plotted for M = 3 (marked as +) and M = 6 (marked as x) from 0 to π.

In the EasyLanguage code of Omega Research's TradeStation2000i, the program for calculating the skipped convolution function of an exponential moving average can be written as follows: -

```
{*****************************************************
Description: Exponential Moving Average skipped convolution function
*****************************************************}

Inputs: Price(NumericSeries), Length(NumericSimple), D(Numeric);
Variables: alpha(0);

alpha = 2 / (Length + 1);
XAverageD = alpha * Price + (1 - alpha) * XAverageD[D];
```

In the above program, alpha is the parameter corresponding to Eq (9.8), and XaverageD is the function corresponding to Eq (9.7).

The program for calculating the indicator for plotting the skipped convolution function of an exponential moving average can be written as follows: -

```
{*****************************************************
Description    : This Indicator plots skipped Exponential Moving Average
*****************************************************}

Inputs: Length(6), D(5);

    Plot1(XAverageD(c, Length, D), "XAverageD");
```

In the above program, the default value of length is taken to be 6, and the default value of the skip parameter, D, is taken to be 5. These values, of course, can be easily changed before execution. The first variable in the function XaverageD, c, corresponds to the Price data in the function program.

An S&P 500 Index data is plotted in Fig 9.3 together with an exponential moving average with M = 6 (thin line), and two skipped exponential moving average with M = 6 and D = 2 (middle thick line) and D = 3 (thickest line). It can be noted that the line with D = 2 is less smooth than the ordinary exponential moving average and the line with D = 3 is less smooth than the line with D = 2. This is consistent with what is described earlier, and also shown in Fig 9.2. The skipped convolution can pick up frequencies larger than π from the input signal. This is a limitation of the skipped convolution. This limitation is related to that of downsampled signal. The relation between skipped convolution and downsampled signal will be considered in the next section.

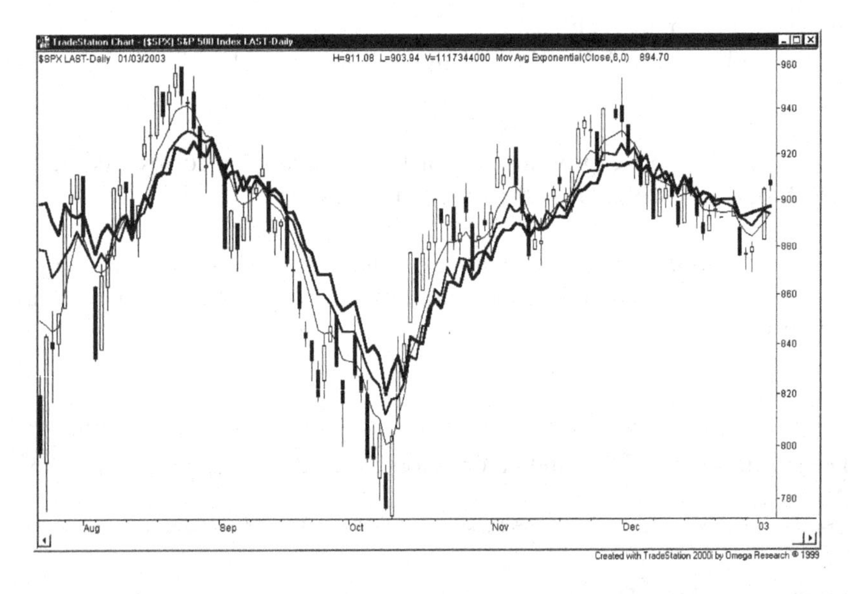

Fig 9.3 An S&P 500 Index data is plotted together with an exponential moving average with M = 6 (thin line), and two skipped exponential moving average with M = 6 and D = 2 (middle thick line) and D = 3 (thickest line). *Chart produced with Omega Research TradeStation2000i.*

9.3 Skipped Convolution and Downsampled Signal

Traders quite often look at financial data in multiple timeframes [Elder 1993]. For example, they will analyze the data in a daily chart, as well as a weekly chart. They would consider the market movements in the weekly chart as the long-term trend, and the price data in the daily chart as short-term actions. The price data in a weekly chart can be considered as taking every fifth data point from a daily chart, but at the end of the week. Thus, the weekly chart can be considered as a downsampled signal from the daily chart. However, there is no reason why a weekly chart cannot be constructed from the daily chart by taking the daily closing price every Thursday or Tuesday. This is why the concept of skipped convolution can be useful. As said earlier, the skipped convolution can alert a trader earlier of a market action.

A signal of frequency $\pi/10$ radian in a daily chart would become a signal of frequency $\pi/2$ in the weekly chart. However, a signal of frequency $\pi/2$ in the weekly chart can also contain other frequencies from the daily chart, namely,

$$\pi/10 + 2\pi/5,\ \pi/10 + 4\pi/5,\ \pi/10 + 6\pi/5,\ \pi/10 + 8\pi/5,$$

which are called aliases [Mak 2003].

A downsampled signal v(n), can be written as

$$v(n) = x(Dn) \tag{9.15}$$

where x(n) is the original signal, and D is the downsampled parameter. That is, every Dth sample of x is taken to construct v.

There exists a relationship between $Y_D(\omega)$, the output response of a skipped convolution of an indicator on the original signal and $H(\omega)V(\omega)$, the output response of an indicator of a downsampled signal. From Eq (9.5) and (9.6),

$$Y_D(\omega) = H_D(\omega)\ X(\omega) = H(D\omega)\ X(\omega) \tag{9.16}$$

Eq (9.16) can be re-written as

$$Y_D(\omega/D) = H(\omega)\ X(\omega/D) \tag{9.17}$$

The Fourier Transform of the convolution between a unit impulse response, h(n) and a downsampled signal v(n) can be written as $H(\omega)V(\omega)$, where $V(\omega)$ is the Fourier Transform of v(n). $V(\omega)$ contains a mixture of frequencies from the original signal. From Mak [2003],

$$H(\omega)V(\omega)$$

$$=\frac{H(\omega)}{D}\left[X\left(\frac{\omega}{D}\right)+X\left(\frac{\omega+2\pi}{D}\right)+\ldots+X\left(\frac{\omega+(D-1)2\pi}{D}\right)\right] \quad (9.18)$$

Using Eq (9.17), Eq (9.18) can be re-written as

$$H(\omega)V(\omega)$$

$$=\frac{1}{D}\left[Y_D\left(\frac{\omega}{D}\right)+Y_D\left(\frac{\omega+2\pi}{D}\right)+\ldots+Y_D\left(\frac{\omega+(D-1)2\pi}{D}\right)\right] \quad (9.19)$$

In deriving Eq (9.19), the following relation is being used:

$$H(\omega+2D\pi)=H(\omega) \quad (9.20)$$

Eq (9.19) thus establishes the relationship between $H(\omega)V(\omega)$, the output response of an indicator on a downsampled signal and $Y_D(\omega)$, the output response of a skipped convolution of an indicator on the original signal. This implies that an indicator applied on a downsampled signal contains more frequencies than a trader would like to have. This definitely would affect the decision made by traders. Further discussions will be given in Chapter 11.

Chapter 10

Trading Tactics

Traders employ different tactics to trade the market. One example is the divergence between price and velocity. They say that that when divergence occurs, the market is going to change direction. However, no explanation is given why this would happen. We will explain this in the next section. Furthermore, they never analyze the tools that they use, and clarify their advantages and limitations. We will examine their popular indicators in the following sections.

10.1 Velocity Divergence

When a piece of stone is thrown upwards, it rises higher and higher, but its speed is getting lesser and lesser. At the highest point, its speed is zero. It then reverses direction. A similar phenomenon appears also in the financial market. While the price is getting higher, the slope of the price is getting smaller. This kind of divergence is exploited by traders, who look at this as good trading opportunities [Pring 1991, Elder 1993, Mak 2003].

Conventionally, slope is represented by momentum (which is actually a two-point moving difference) by traders. However, as is pointed out by Mak [2003], the newly coined velocity indicators have much less phase lag than momentum. Velocity will thus be used instead of momentum here to analyze the data.

Thus, if there is a divergence between the price and the velocity, it is forecasted that the price will change direction. An example is given in Fig 10.1, where the S&P500 weekly data in the year 2002 and 2003 is plotted. The S&P 500 reached an all time high of approximately 1550 in March 2000. It then dropped 50% to approximately 775, forming a triple bottom in July, October 2002 and March 2003. In the price chart in Fig 10.1, an adaptive moving average (created by Jurik Research) of smoothness 3 is plotted as a line. A quartic velocity and a quartic acceleration of the adaptive moving average are plotted in the middle and bottom figures respectively. Slightly before the three points where the S&P 500 index are hitting triple bottoms, the successive magnitudes of the (negative) quartic velocity are getting smaller and smaller, thus showing a divergence (Note that the velocity of price leads the price). The implication is that the index would soon rise. As it happened, it did rise like a phoenix. The velocity maintains more or less positive after April, 2003, in consistence with the rising price. The advantage about price and velocity divergence is that this occurrence precedes the turning point, where the velocity is zero, thus giving plenty of time to alert the trader.

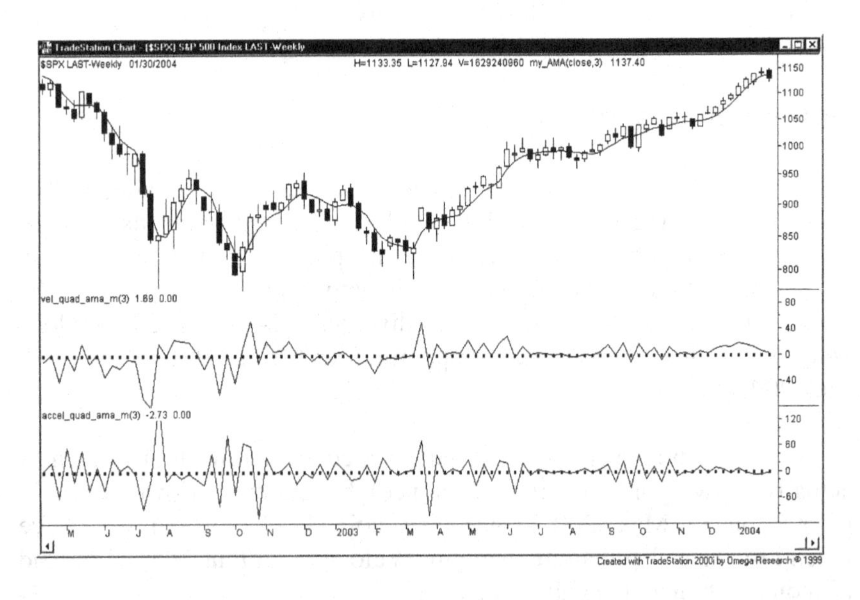

Fig 10.1 A weekly chart of the S&P 500 Index. The price data was smoothed using the adaptive moving average with smoothness 3 (shown as a line in the top figure). Quartic velocity and quartic acceleration indicators are plotted in the middle and bottom figures respectively. *Chart produced with Omega Research TradeStation2000i.*

10.2 Moving Average Convergence-Divergence (MACD)

10.2.1 *MACD Indicator*

The MACD indicator is a popular indicator used by traders [Elder 1993, 2002; Pring 1991; Appel 1991]. The indicator was developed by Gerald Appel, an analyst and money manager in New York [Elder 1993, 2002].

The indicator consists of two lines: the MACD line and the Signal line. The MACD line is composed of two exponential moving averages, a fast EMA and a slow EMA. The line is called MACD (moving average convergence-divergence) because the fast EMA is continually converging toward and diverging from the slow EMA. Fast EMA has a smaller length, M, than that of slow EMA. The slow EMA (e.g. $M_2 = 26$) is subtracted from the fast EMA (e.g. $M_1 = 12$). Their difference is plotted. This is called the (fast) MACD line. An EMA (e.g. $M_3 = 9$) of the (fast) MACD line is calculated and plotted on the same chart. This is called the slow Signal line.

The trading rule is based on the crossover between the MACD and Signal lines. When the fast MACD line crosses above the slow Signal line, a buy signal is generated. When the fast line crosses below the slow line, a sell signal is implemented.

This trading rule is somewhat similar to the crossover between a fast EMA line and a slow EMA line [Pring 1991; Mak 2003]. The trader will buy when the fast EMA crosses over and is above the slow EMA. He will sell when the fast EMA crosses over and is below the slow EMA.

Now, the question is: is the MACD trading rule better than the EMA trading rule? And what exactly is the MACD line? We will find that out in the next section.

10.2.2 *MACD Line*

The MACD line is the difference between two exponential moving averages. We would like to find out the amplitude and phase of the Fourier Transform of the MACD line.

The Fourier Transform of an exponential moving average is a low pass filter. The Fourier Transform of the difference of two exponential moving averages is simply the difference in their respective Fourier Transform [Brigham 1974, P31]. Let $H_1(\omega)$ and $H_2(\omega)$ be the Fourier Transform of the fast EMA and the slow EMA respectively. The Fourier Transform of the MACD line, $H_4(\omega)$ is simply given by:

$$H_4(\omega) = H_1(\omega) - H_2(\omega) \tag{10.1}$$

10.2.2.1 *Fast EMA($M_1 = 12$) and Slow EMA($M_2 = 26$)*

Taking the lengths of the fast EMA and the slow EMA to be 12 and 26 respectively (12 and 26 are common parameters used by traders [Elder 1993, 2002]), their amplitudes as well as the amplitude of $H_4(\omega)$ are plotted in Fig 10.2(a). This simply shows that $H_4(\omega)$ is a band-pass filter. When the amplitudes of two low pass filters are the same at $\omega = 0$, their difference should yield a band-pass filter. The MACD line filters off the very slow moving market movement near $\omega = 0$, thus removing some of the slowly changing trend. This can make the timing of a crossover between an MACD line and the Signal line a better timing than the crossover of a fast EMA line and a slow EMA line. Fig 10.2(a) also shows that the MACD line in this particular case is as smooth as the slow EMA, making it a very good filter.

To show that the MACD line function even better than the EMA, we can take a look at Fig 10.2(b), which plots the phases of the fast EMA and slow EMA, as well as the phase of the MACD line. It can be seen that the MACD line has a lesser phase lag than even the fast EMA. At $\omega = 0$, the real and imaginary part of $H_4(\omega)$ is 0. Thus, the phase of the MACD line at $\omega = 0$ is of the indeterminate form. However, using the De l'Hopital's Rule [e.g. Kaplan 1952, P27], it can be shown that the phase is equal to $\pi/2$. The phase $\pi/2$ is thus plotted at $\omega = 0$ for $H_4(\omega)$. Therefore, because the MACD line has a smaller phase lag, the MACD trading technique would provide a quicker crossover and therefore buy or sell signal than a comparative EMA trading technique. Thus, it is no wonder that some experienced traders embrace the MACD trading tactic.

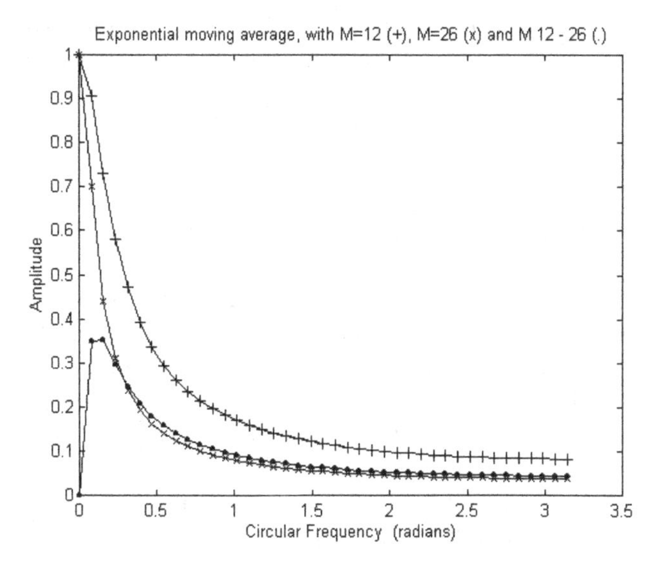

Fig 10.2(a) The amplitudes of the Fourier Transform of the fast EMA ($M_1 = 12$) (plotted as +) and the slow EMA ($M_2 = 26$) (plotted as x), and the amplitude of the MACD line (plotted as .) are plotted versus the circular frequency ω.

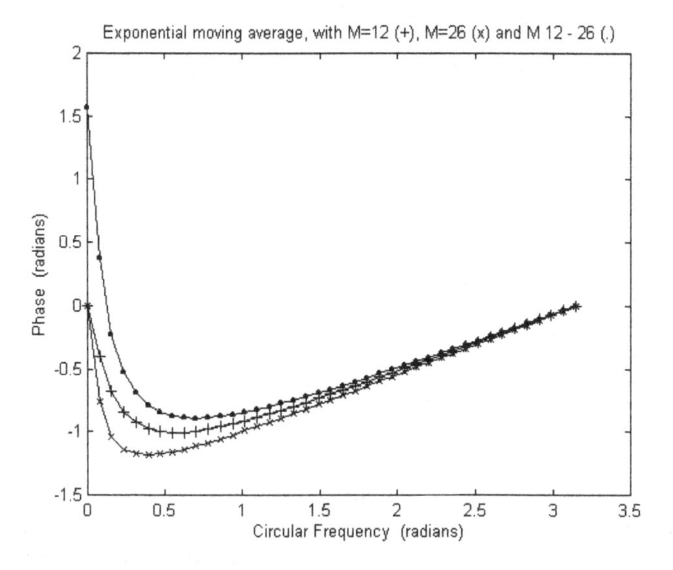

Fig 10.2(b) The phases of the Fourier Transform of the fast EMA ($M_1 = 12$) (plotted as +) and the slow EMA ($M_2 = 26$) (plotted as x), and the phase of the MACD line (plotted as .) are plotted versus the circular frequency ω.

An example would be given. In Fig 10.3, the top figure plots a signal, p, composed of waves of two frequencies to simulate market price data:

$$p = 0.25\sin(\pi n/10) + \sin(\pi n/40) \tag{10.2}$$

where n is an integer.

Two exponential moving averages (EMA) of length 12 and 26 are plotted on the same figure. Their crossovers would provide entry and exit points for traders to trade the market. The bottom figure in Fig 10.3 plots the MACD line (M_1 = 12 and M_2 = 26) and the Signal line (M_3 = 9). Again, their crossover would provide entry and exit points for trading the market. It can be seen that the crossovers of the MACD line and Signal line provides much better timing and therefore profitability than the crossovers of the two EMA lines.

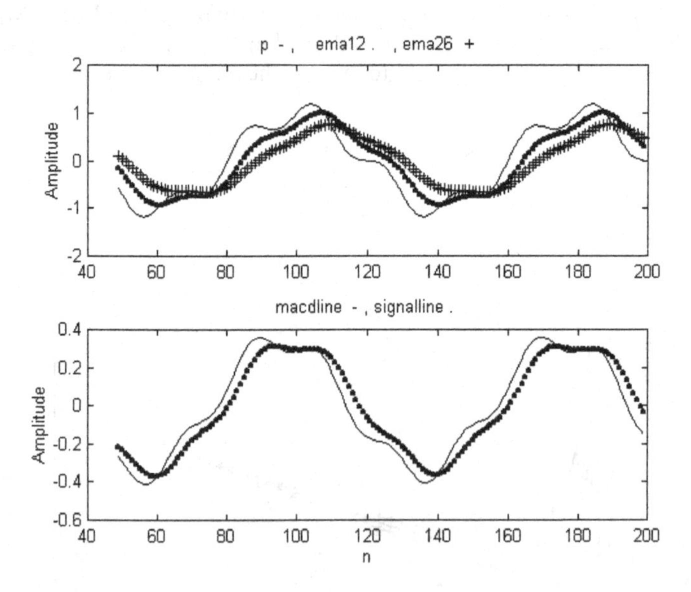

Fig 10.3 The top figure plots a simulated price data as a line. The fast EMA (M_1 = 12) (plotted as .) and the slow EMA (M_2 = 26) (plotted as +) of the price data are also plotted. The bottom figure plots the MACD line (plotted as a line) and the Signal line (plotted as .). Note that the crossovers of the MACD line and the Signal Line appear earlier than the crossovers of the two EMA lines, thus providing better timing for trading purposes.

10.2.2.2 *Fast EMA($M_1 = 5$) and Slow EMA($M_2 = 34$)*

The lengths of the fast and slow EMA for the MACD line are quite often taken to be 5 and 34 [Elder 1993]. The amplitudes and phases of the two EMA's and their MACD line are plotted in Fig 10.4(a) and (b) respectively. Comparing Fig 10.4(a) with 10.2(a), it can be seen the amplitudes of both MACD lines peak at about 0.2 radian. Fig 10.4(b) shows that, again, the MACD line has a lesser phase lag than even the fast EMA. However, Fig 10.4(a) shows that it is only about as smooth as the fast EMA. Thus, it is not as good a band-pass filter as the one we discussed in section 10.2.2.1. This simply means that the length of the EMA's have to be chosen carefully to optimize one's trading tactics. In order that the MACD line is approximately as smooth as the slow EMA, the length of the slow EMA should be approximately twice as long as that of the fast EMA. Another set of lengths commonly used for the fast and slow EMA's are 8 and 17 [Pring 1991].

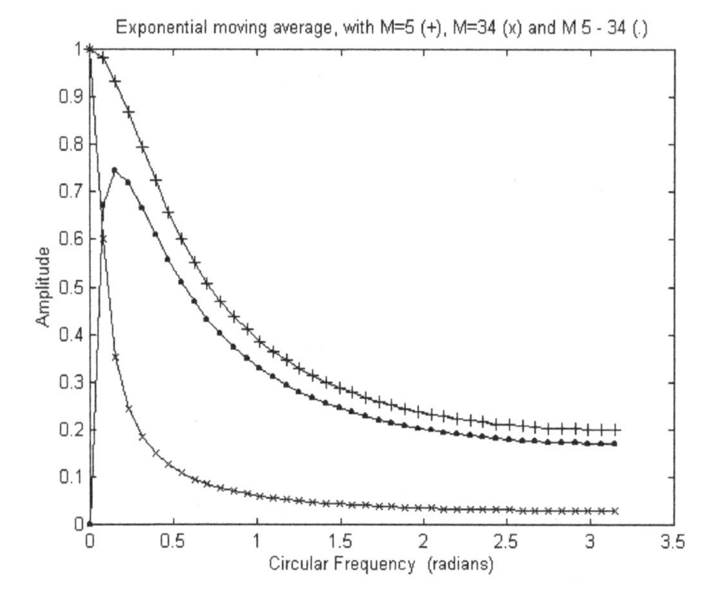

Fig 10.4(a) The amplitudes of the Fourier Transform of the fast EMA ($M_1 = 5$) (plotted as +) and the slow EMA ($M_2 = 34$) (plotted as x), and the amplitude of the MACD line (plotted as .) are plotted versus the circular frequency ω.

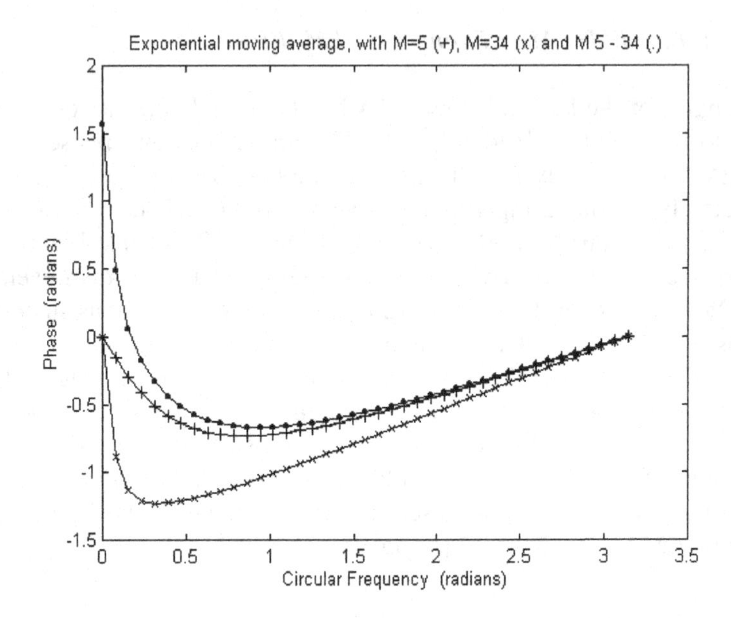

Fig 10.4(b) The phases of the Fourier Transform of the fast EMA (M_1 = 5) (plotted as +) and the slow EMA (M_2 = 34) (plotted as x), and the phase of the MACD line (plotted as .) are plotted versus the circular frequency ω.

10.3 MACD-Histogram

It has been said that the MACD-Histogram offers a better insight into the balance of power between the bulls and bears than the MACD [Elder 1993, 2002]. Not only does it show whether the bulls or bears are in control, it also shows whether they are growing stronger or weaker. Now, are these claims correct? First, let us see how the MACD-Histogram is defined. It is defined as the difference between the MACD line and the Signal line:

$$\text{MACD-Histogram} = \text{MACD line} - \text{Signal line} \qquad (10.3)$$

The MACD-Histogram is usually plotted as a histogram to distinguish it from the MACD line, but it can just as well be plotted as a line. As discussed earlier, the signal line is the exponential moving average of the MACD line. Let the Fourier Transform of this exponential moving average be denoted by $H_3(\omega)$. Then the Fourier Transform of the Signal line will be given by $H_3(\omega)\ H_4(\omega)$. The Fourier Transform of the MACD-Histogram, $H_5(\omega)$, according to Eq (10.3) will be given by

$$H_5(\omega) = H_4(\omega) - H_3(\omega)\ H_4(\omega) \tag{10.4a}$$

$$= (1 - H_3(\omega))H_4(\omega) \tag{10.4b}$$

As $H_3(\omega)$ is a low pass filter, and $H_4(\omega)$ is a band-pass filter, $H_3(\omega)\ H_4(\omega)$ is a band-pass filter. Eq (10.4a) would mean that $H_5(\omega)$ is a band-pass filter, which is formed from the difference between two band-pass filters. Eq (10.4a) can be rewritten in the form of Eq (10.4b). As $(1 - H_3(\omega))$ is a high pass filter, $H_5(\omega)$ can also be considered as a combination between a high pass filter and a band-pass filter.

Taking the lengths of the fast EMA, slow EMA and the EMA of the MACD line to be 12, 26 and 9, the amplitudes and phases of the MACD line, Signal line and MACD-Histogram are plotted in Fig 10.5(a) and (b). Fig 10.5(a) shows that the MACD-Histogram is a band-pass filter, having a peak at a frequency slightly larger than that of the MACD line (which is also a band-pass filter). It filters off more of the low frequency component of a signal than the MACD line. Fig 10.5(b) shows that it has less phase lag than the MACD line, thus the MACD-Histogram can detect market movement faster. The phase of the Signal line is the sum of the phase of $H_3(\omega)$ and the phase of $H_4(\omega)$. At $\omega = 0$, since the phase of $H_3(\omega) = 0$ and the phase of $H_4(\omega) = \pi/2$, the phase of $H_3(\omega) = \pi/2$. At $\omega = 0$, the amplitude of $H_5(\omega)$ is 0, which implies that the real and imaginary part of $H_5(\omega)$ is 0, and the phase is in an indeterminate form. To calculate that phase, it would be easier to use Eq (10.4b) than (10.4a), as the phase of $H_5(\omega)$ is simply given by the sum of the phase of $(1 - H_3(\omega))$and the phase of $H_4(\omega)$. Using again the De l'Hopital's Rule, the phase of the high pass filter, $(1 - H_3(\omega))$ at $\omega = 0$ is found to be $\pi/2$. As the phase of $H_4(\omega)$ at $\omega = 0$ is $\pi/2$, the phase of $H_5(\omega)$ is π.

The lengths of the EMA associated with the MACD, 12, 26 and 9 have become standard values used as defaults in most trading software packages. It has been said that changing these numbers has little impact on the MACD signals, unless their ratios are changed drastically [Elder 2002]. This is probably due to the fact that the peak of the MACD-Histogram does not change much even though those numbers are changed. The peak of the MACD Histogram for the values 12, 26, and 9 is located at 0.21 radians (Fig 10.5(a)). If the values were changed to 3, 6, and 9, the peak would only shift to 0.58 radian (Fig 10.6(a)). However, this slight shift can make a difference, which is more accentuated in the phase plot. Fig 10.6(b) plots the phase of the MACD line, Signal line and MACD-Histogram for the values 3, 6, and 9. It should be noted that the phase of the MACD-Histogram is almost zero for circular frequencies greater than 0.5 radian. As its phase lag is less than that of the MACD-Histogram with values of 12, 26 and 9, this property seems to make it a good candidate as a band-pass filter. However, its usefulness is much limited as it retains a certain portion of high frequency noises.

For years, traders have been attempting to develop zero lag low pass filters without too much success [Ehlers 1992, 2001]. But the trick is to develop not low pass filter, but band pass filter with almost zero phase lag. MACD line and MACD Histogram are band pass filters with almost zero lag. Their characteristics have not been pointed out so far. Thus, it does need mathematical analysis to discover the importance of these filters.

Qualitatively, the MACD-Histogram does not provide any significant difference from the MACD line, as contrasted to what some traders would like to think. However, the MACD-Histogram does filter out more low frequency component than the MACD line. This can make it a very useful indicator. While an exponential moving average (e.g. an EMA of length = 26) can monitor the long term trend of the market, the MACD-Histogram can denote the middle term movement with very little time lag. High frequency components are usually eliminated in trading, as they are considered as noise. Thus, as long as the trader keeps an eye on the long and middle trends, he would have a good understanding of the market.

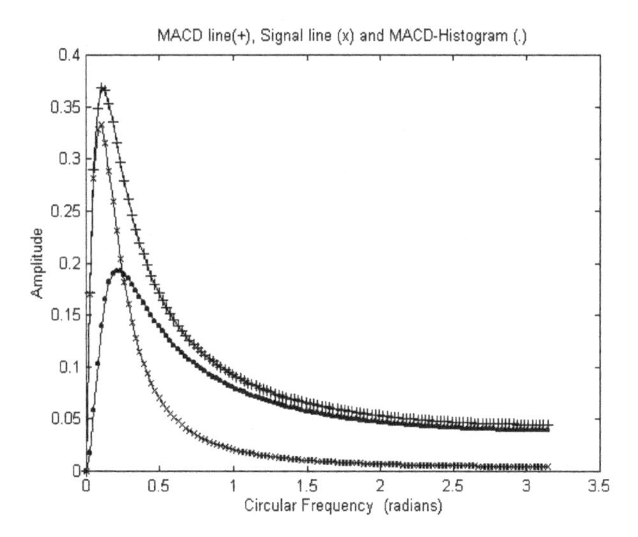

Fig 10.5(a) The amplitudes of the MACD line (plotted as +), Signal line (plotted as x) and MACD-Histogram (plotted as .) are plotted versus the circular frequency ω. The lengths of the fast EMA, slow EMA and the EMA of the MACD line are taken to be 12, 26 and 9 respectively.

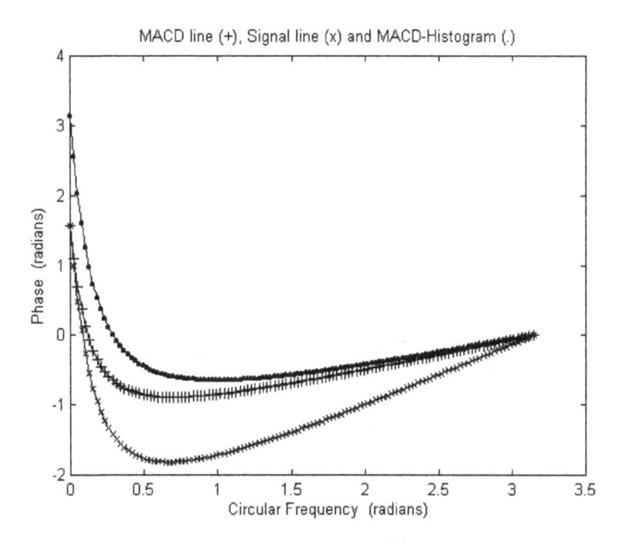

Fig 10.5(b) The phases of the MACD line (plotted as +), Signal line (plotted as x) and MACD-Histogram (plotted as .) are plotted versus the circular frequency ω. The lengths of the fast EMA, slow EMA and the EMA of the MACD line are taken to be 12, 26 and 9 respectively.

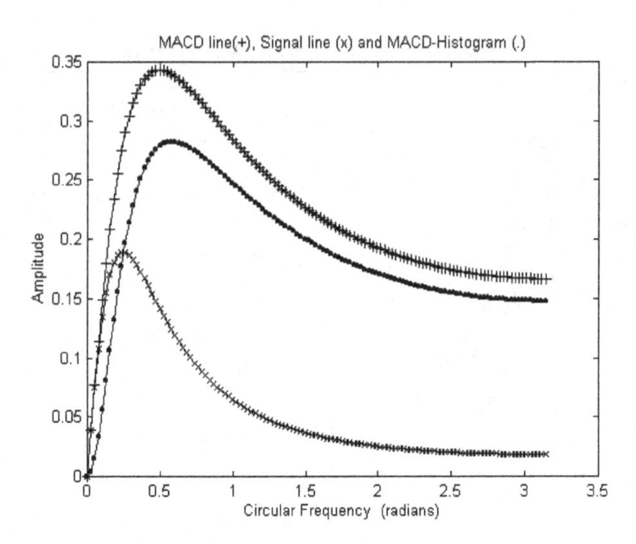

Fig 10.6(a) The amplitudes of the MACD line (plotted as +), Signal line (plotted as x) and MACD-Histogram (plotted as .) are plotted versus the circular frequency ω. The lengths of the fast EMA, slow EMA and the EMA of the MACD line are taken to be 3, 6 and 9 respectively.

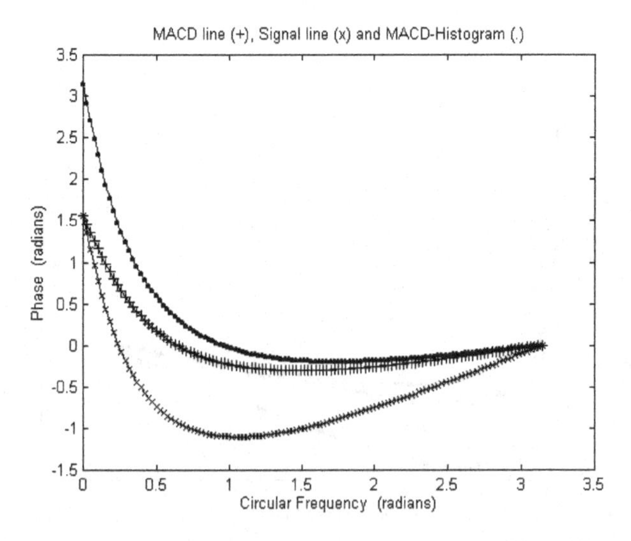

Fig 10.6(b) The phases of the MACD line (plotted as +), Signal line (plotted as x) and MACD-Histogram (plotted as .) are plotted versus the circular frequency ω. The lengths of the fast EMA, slow EMA and the EMA of the MACD line are taken to be 3, 6 and 9 respectively.

10.3.1 *MACD-Histogram Divergence*

Some traders claim that the divergences between MACD-Histogram and price give some of the strongest signals in technical analysis. These divergences identify major turning points and yield strong buy or sell signals [Elder 1993, 2002]. When prices fall to a new low but the indicator falls to a more shallow low than the preceding one, the MACD-Histogram traces a bullish divergence. This implies that the bears have grown weaker and traders should go long. When prices rise to a new high but the indicator rises to a lower peak than the preceding one, the MACD-Histogram traces a bearish divergence. The bulls are receding and traders should go short. These deductions are actually not quite true.

It should be noted that there is a fundamental difference between MACD-Histogram divergence and velocity (or momentum) divergence. Velocity (or momentum) is a high-pass filter that simulates the slope of the price data [Mak 2003]. Thus, divergence between price and velocity implies that price and its slope are going into different directions. The market is going to turn. The MACD-Histogram, as discussed above, is a band-pass filter. It filters off the low frequency (long cycles) and high frequency (short cycles), leaving behind the middle frequency. Its divergence from the price means that its middle frequency signal generates a different opinion from that of the total price.

If the lengths of the exponential moving averages are chosen to be $M_1 = 12$, $M_2 = 26$ and $M_3 = 9$, the MACD-Histogram will peak at about a circular frequency of approximately 0.21 radians (see Fig 10.5(a)). We will see how an MACD-Histogram divergence can arise.

As an example, we will simulate the market price movement as follows:

$$p = 0.25\sin(\pi n/10) + \sin(\pi n/40) \qquad (10.5)$$

where p is the price,

and n is a positive integer.

The frequency of $\pi/10$ radian corresponds to a $2\pi/(\pi/10) = 20$ bar cycle. (If a daily chart is being used, a 20 bar cycle means that the cycle is approximately one month.) In Fig 10.7, p is plotted in the top figure, while the MACD-Histogram of p is plotted in the bottom figure. The MACD-Histogram, acting as a band-pass filter, filters out part of the $\pi/40$ signal and retains most of the $\pi/10$ signal. As it can be observed in the figure at n = 85 and n = 105, while the price, p, rises to a new high, the Histogram rises to a lower peak. Trading rule would dictate the trader to go short. As it happens, the price, p, does turn down. At n = 115 and n = 135, the total price, p, falls to a new low, the Histogram falls to a higher low. Trading rule would dictate the trader to go long. The price, p, does turn up after. Thus, in this particular example, the trading rule of MACD-Histogram divergence works very well.

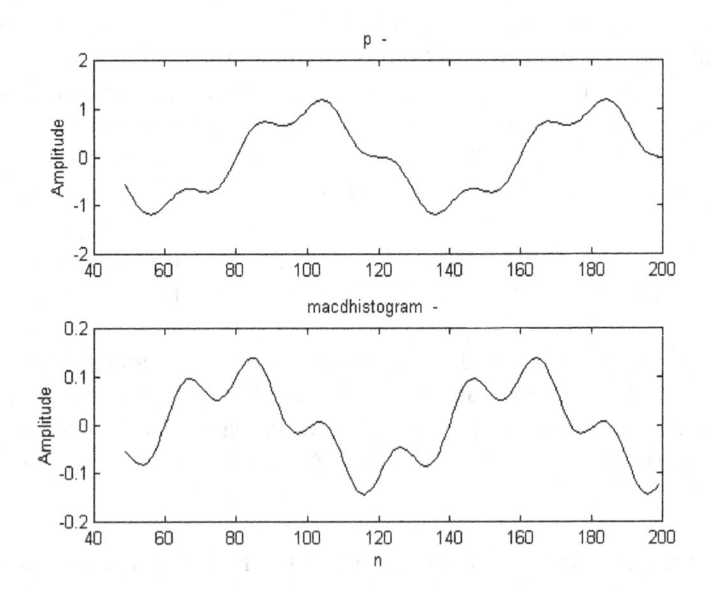

Fig 10.7 The top figure plots a simulated price $p = 0.25\sin(\pi n/10) + \sin(\pi n/40)$. The bottom figure plots the macdhistogram of the price. The trading rule of MACD-Histogram divergence works in this case.

We will take a look at another example:

$$p = 0.25\sin(\pi n/5) + \sin(\pi n/20) \qquad (10.6)$$

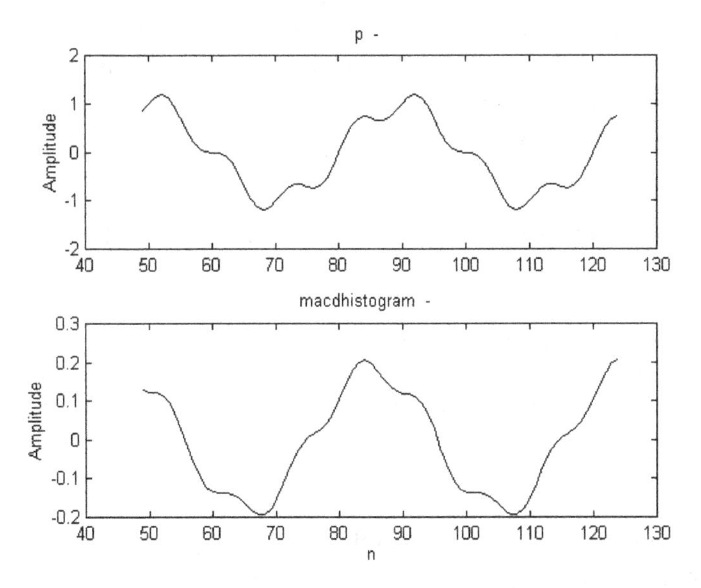

Fig 10.8 The top figure plots a simulated price $p = 0.25\sin(\pi n/5) + \sin(\pi n/20)$. The bottom figure plots the macdhistogram of the price. The trading rule of MACD-Histogram divergence does not necessarily work.

In Fig 10.8, p is plotted in the top figure, while the MACD-Histogram of p is plotted in the bottom figure. As it can be observed in the figure at $n = 84$ and $n = 92$, while the price, p, rises to a new high, the Histogram rises to a lower peak. This shows a bearish divergence. As it happens, the price, p, does turn down. At $n = 98$ and $n = 108$, the total price, p, falls to a new low, the Histogram also falls to a new low. No divergence is thus observed. However, the price, p, turns up after.

Thus, the occurrence of any MACD-Histogram divergence would depend very much on the frequencies of the market price signal. The divergence may not happen sometimes. Hence, it is believed that the velocity divergence would be a more reliable indicator. The velocity divergence has also an advantage over the MACD-Histogram divergence. It can be observed in Fig 10.7 and 10.8 that the peaks (or valleys) of the MACD-Histogram do not have any phase lead as compared to the peaks (or valleys) of the price. This, of course, is consistent with the phase plot in Fig 10.5(b). However, the peaks (or valleys) of the velocity of the price has a phase lead of approximately $\pi/2$ compared to the peaks (or valleys) of the price [Mak 2003]. This provides an early warning of the market turn, which would happen when the velocity becomes zero later.

10.4 Exponential Moving Average of an Exponential Moving Average

Exponential moving average (EMA) is quite often used to smooth the price data. The direction of its slope is considered to be an important message. When the EMA rises, the crowd is bullish. It is a good time to go long. When the EMA falls, the crowd is bearish. It is a good time to go short [Elder 2002].

An EMA with a short length is quite sensitive to price changes. It allows the trader to catch new trends sooner. However, it also changes its direction more often. An EMA with a longer length does not change direction so often, but it has a larger time lag in recognizing turning points. As a rule, the longer the trend the trader attempts to catch, the larger the length of the EMA should be chosen.

The longer the length of the EMA, the smoother the data. Alternatively, some traders suggest smoothing the EMA with another EMA [Pring 1991]. However, this would create a much larger time lag with respect to the data.

Smoothing one EMA with another EMA is equivalent to convoluting the unit impulse response of the EMA with the unit impulse response of another EMA. The Fourier Transform of the operation is equal to the product of their individual Fourier Transform [Brigham 1974]. The phase is therefore equal to the sum of their individual phases.

As an example, an EMA of length 6 is applied onto an EMA of length 3. The amplitude and phase of its Fourier Transform is shown in Fig 10.9(a) and (b) respectively. The amplitude is equal to the product of the amplitude of an EMA of length 6 and the amplitude of an EMA of length 3. The phase is equal to the sum of the phase of an EMA of

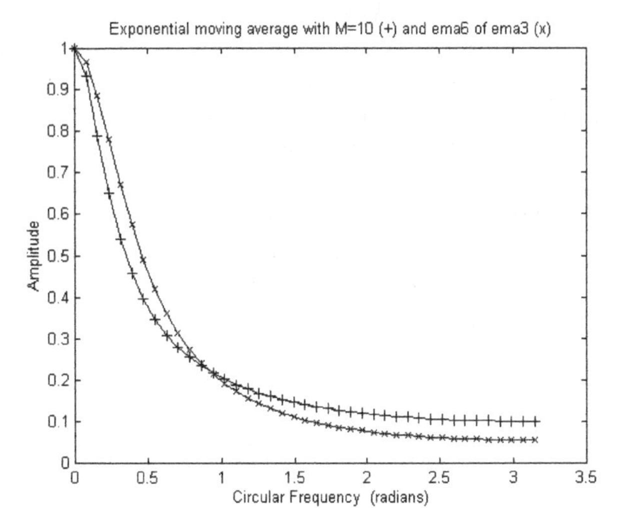

Fig 10.9(a) The amplitude of the Fourier Transform of an EMA of length 6 applied onto an EMA of length 3 is plotted (as x) versus the circular frequency ω. The amplitude of the Fourier Transform of an EMA of length 10 is also plotted (as +) for comparison purpose.

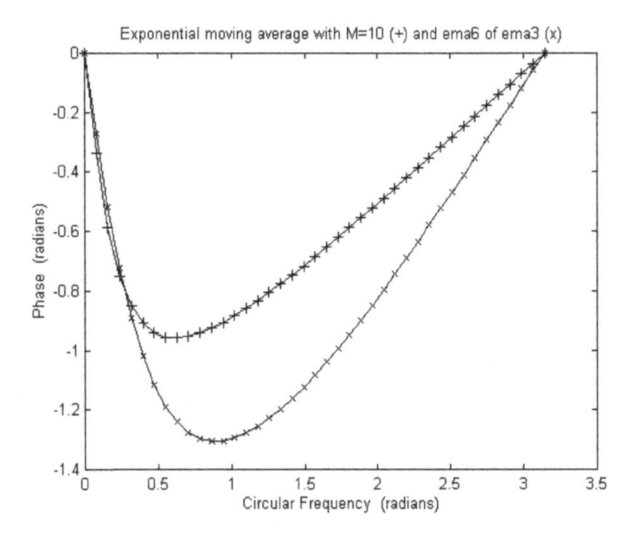

Fig 10.9(b) The phase of the Fourier Transform of an EMA of length 6 applied onto an EMA of length 3 is plotted (as x) versus the circular frequency ω. The phase of the Fourier Transform of an EMA of length 10 is also plotted (as +) for comparison purpose.

length 6 and the phase of an EMA of length 3. The amplitude and phase of the Fourier Transform of an EMA of length 10 are also plotted in the same figures for comparison purpose. It can be seen from Fig 10.9(a) that it has a similar filtering capability as the EMA of an EMA. The EMA of the EMA does produce a smoother line, as it retains more low frequency component and rejects more high frequency component (Fig 10.9(a)). However, this is done at the expense of a much larger phase lag (Fig 10.9(b)). Thus, it would seem that it would be more practical to choose an EMA of longer length to smooth the data than to use an EMA of an EMA.

Chapter 11

Trading System

Many trading systems have been proposed by traders. However, their usefulness is seldom analyzed. In this chapter, we will discuss how we can analyze a trading system mathematically. Some trading systems use multiple timeframes. We will discuss the advantages and disadvantages. One trading system, the triple screen trading system, will be taken as an example

11.1 Multiple Timeframes

Some traders recommend using several timeframes to analyze the market. A method of using triple screens has been suggested by Elder [1993, 2002]. In the first screen, a strategic decision to trade long or short is made using a trend-following indicator on a long-term chart. In the second screen, a tactical decision about entries and exits is formed using oscillators on an intermediate-term chart. (Oscillators measure the speed or slope of the trend. Examples of oscillators are momentum, velocity, etc.) In the third screen, methods for placing buy and sell orders are implemented on a short-term chart, or on the same intermediate-term chart. The method can consist of, e.g., using a breakout or pullback to enter trades.

An intermediate timeframe is any timeframe chosen by the trader. It can be a weekly chart, a daily chart, or a five-minute chart. The long-term chart is decided by multiplying the unit time interval of the intermediate chart by a certain factor, e.g., five. If the intermediate chart is daily, then the long-term chart is weekly. If the intermediate chart is five minutes, then the long-term chart can be half-hourly. Similarly, the short-term chart is decided by dividing the unit time interval of the intermediate chart by the same factor. For a daily intermediate chart, the short-term chart can be hourly. For a five-minute chart, the short-term chart is a one-minute chart. The factor chosen is not that critical. It can be a factor of four instead of five. Thus, if a trader's intermediate timeframe is a weekly chart, his long-term frame will be a monthly chart.

What does it mean by going from an intermediate timeframe to a long-term timeframe? Let's take the factor of five. It simply means taking every fifth point of the data points in the intermediate time frame to form the data points in the long-term frame. Mathematically, it is called downsampling five [Mak 2003]. An original signal of frequency $\pi/10$ radian will become a signal of $\pi/2$ radian in the long-term frame. A signal of $\pi/50$ radian will become a signal of $\pi/10$ radian. You can imagine the frequencies in the frequency response plot of the market price signal in the intermediate timeframe to stretch by a factor of five to form the frequencies in the frequency response plot in the long-term frame. Does a long-term frame help in analyzing the market? We will find out in the following section.

11.1.1 *Long-Term Timeframe*

11.1.1.1 *Advantages*

It is common for traders to apply a trending indicator, e.g., an exponential moving average (EMA) on the data. As said earlier, a trending indicator is a low pass filter. Compared to an exponential moving average of a certain length applied to the data in an intermediate time-chart, downsampling the data to a long-term time chart and applying the same exponential moving average would mean pushing more signals of high frequencies out of the original signal in the intermediate time frame. Signals of low frequencies will be more emphasized. This effectively means increasing the length of the exponential moving average in the intermediate time chart, thus narrowing the bandwidth of the low pass filter. However, there is an advantage of downsampling to a long-term chart as compared to increasing the length of an EMA in an intermediate chart. There is less phase lag.

Mathematically, we would not be able to calculate the frequency response of an indicator on downsampled signal. However, we can calculate the frequency response of a skipped convolution of an indicator, which should provide some insight to an indicator on downsampled signal.

Fig 11.1(a) plots the amplitude response of an exponential moving average of length 130, and the amplitude response of a skipped 5 convolution of an exponential moving average of length 26. It can be seen that their amplitudes are approximately the same for $\omega < 0.5$. That means, for $\omega < 0.5$, the two EMA's have similar smoothing capability. It can actually be shown that, for ω close to 0, a skipped D convolution of an exponential moving average of length M has similar smoothing capability to an exponential moving average of length approximately equal to DM, where D is the skip parameter.

Fig 11.1(b) plots the phase response of an exponential moving average of length 130, and the phase response of a skipped 5 convolution of an exponential moving average of length 26. It can be seen that, except in the high frequency range, the skipped convolution has much less phase lag. This implies that the EMA of a downsampled signal has less phase lag than the phase of an EMA of similar smoothing capability in the original signal.

Thus, analyzing market data in a long-term timeframe does have an advantage. The trending indicator would have less phase lag compared to the same trending indicator with similar smoothing capacity in an intermediate timeframe.

The long-term frame has also an obvious advantage that the trader has less data to handle and visualize. He can see the forest instead of the trees. However, it does have its disadvantages.

11.1.1.2 *Disadvantages*

(1) While trending indicators, e.g., the EMA, provides less phase lag in downsampled signal than the original signal, oscillator indicators, e.g., the momentum indicator and velocity indicators provide more phase lag. We will illustrate this with an example in the frequency domain. Fig 11.2(a) shows the amplitude of the frequency response of the cubic

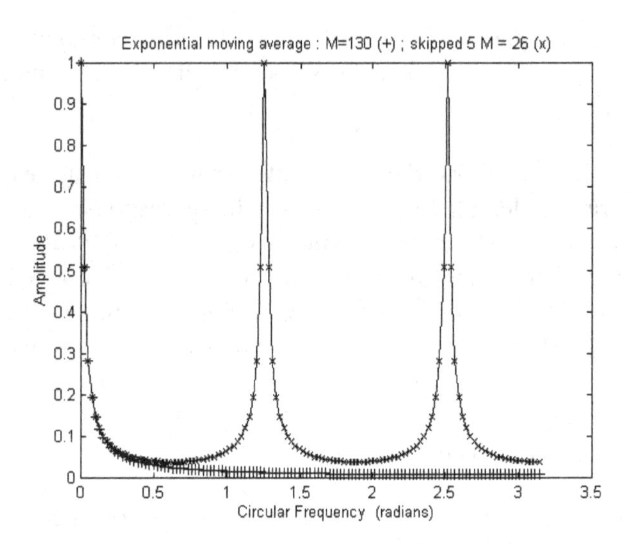

Fig 11.1(a) The amplitudes of the Fourier Transform of an exponential moving average of length 130 (plotted as +) and a skipped 5 convolution of an exponential moving average of length 26 (plotted as x) are plotted versus the circular frequency ω.

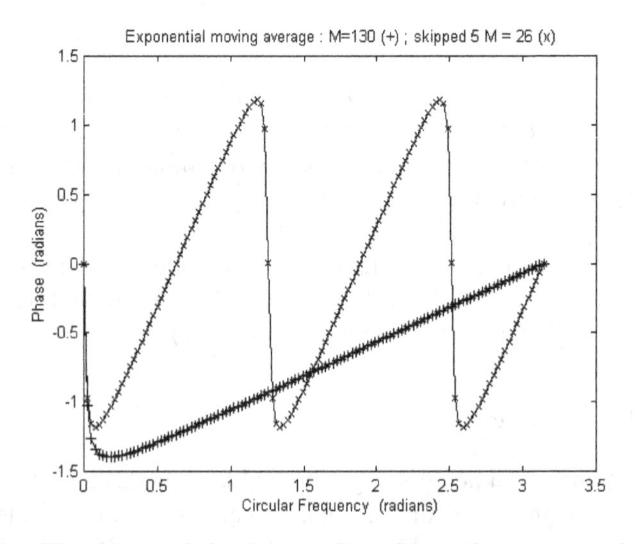

Fig 11.1(b) The phase of the Fourier Transform of an exponential moving average of length 130 (plotted as +) and a skipped 5 convolution of an exponential moving average of length 26 (plotted as x) are plotted versus the circular frequency ω.

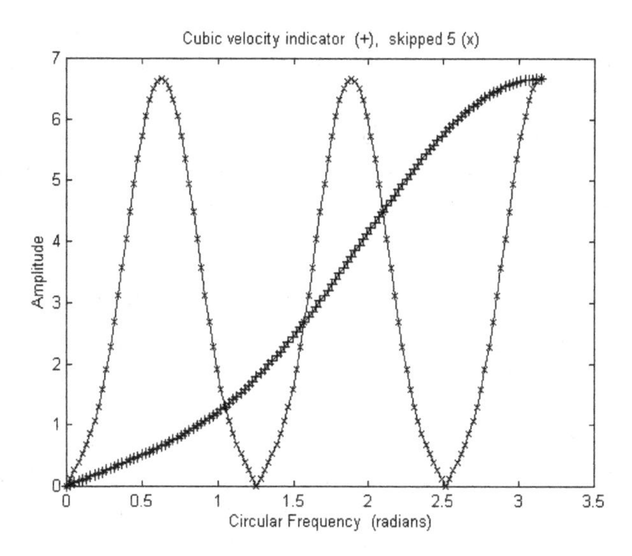

Fig 11.2(a) The amplitudes of the Fourier Transform of the cubic velocity indicator (plotted as +) and the skipped 5 convolution of the cubic velocity indicator (plotted as x) are plotted versus the circular frequency ω.

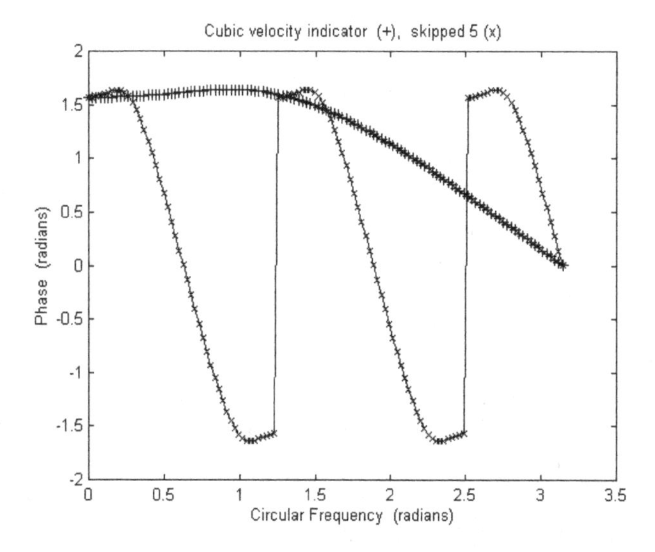

Fig 11.2(b) The phases of the Fourier Transform of the cubic velocity indicator (plotted as +) and the skipped 5 convolution of the cubic velocity indicator (plotted as x) are plotted versus the circular frequency ω.

velocity indicator as well as the skipped 5 convolution of the cubic velocity indicator. Fig 11.2(b) shows the phase of the frequency response of the cubic velocity indicator as well as the skipped 5 convolution of the cubic velocity indicator. From Fig 11.2(b), it can be implied that except for frequency near π, cubic velocity on downsampled signal has a larger phase lag than the cubic velocity on the original signal.

We will also take a look at an example in the time domain. Fig 11.3(a) shows an original price signal having a frequency of $\pi/16$ radian (denoted by .) and the down 4 price signal which has a frequency of $\pi/4$ (denoted by o). The 2-bar momentum indicator (which is the same as the 2-point moving difference) applied to the original signal has a phase lag of $\pi/32$ (= $\pi/16 \times \frac{1}{2}$) radian from the ideal phase lead of $\pi/2$ [Mak 2003]. The output is shown in Fig 11.3(b). The unit impulse response of the momentum indicator is defined here as (1 , –1). As it takes two points to calculate the momentum, the very first point (which is arbitrarily set to zero) in Fig 11.3(b) should be ignored. Fig 11.3(c) shows the down 4 momentum signal. It can be seen that the phase difference between the downsampled momentum and the downsampled price (in Fig 11.3(a)) is preserved. However, if the price signal is downsampled first (thus becoming a signal of frequency of $\pi/4$), and then the momentum indicator is applied to the downsampled price, the output would have a phase lag of $\pi/8$ (= $\pi/4 \times \frac{1}{2}$) radian from the ideal phase lead of $\pi/2$, when compared to the downsampled price signal. Fig 11.4(a) shows the original price data. Fig 11.4(b) shows the down 4 price data. The momentum indicator is applied to the down 4 price data, and plotted in Fig 11.4(c). Again, as it takes two points to calculate the momentum, the very first point (which is arbitrarily set to zero) in Fig 11.4(c) should be ignored. It can be seen that the phase lag of the momentum from the downed price data is much larger in Fig 11.4(c) than that shown in Fig 11.3(c). This is simply caused by the fact that the phase lag of momentum of signal of frequency $\pi/4$ is much larger than that of signal of frequency $\pi/16$. Thus, it can make quite some difference whether the indicator is applied in the long-term time-frame or the indicator is applied in the original time-frame and the output is then downsampled to the long-term time-frame. In other words, the indicator operation and downsampling is not commutative.

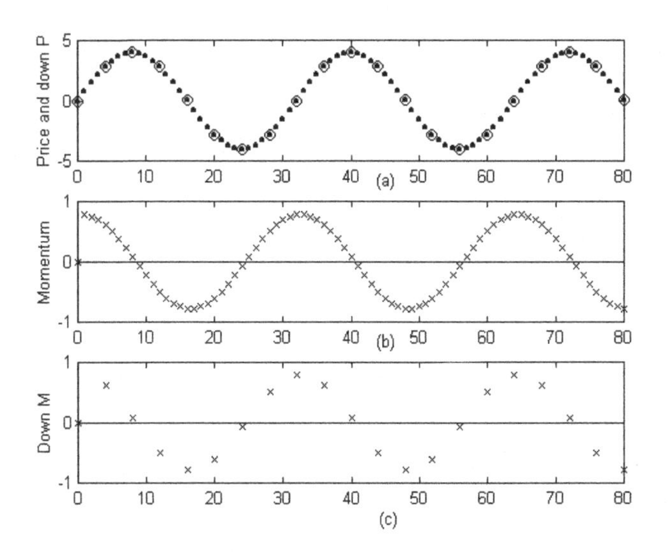

Fig 11.3(a) shows an original price signal having a frequency of $\pi/16$ radian (plotted as .) and the down 4 price signal which has a frequency of $\pi/4$ (plotted as o). (b) shows the 2-bar momentum indicator (which is the same as the 2-point moving difference) applied to the original signal. It has a phase lag of $\pi/32$ (= $\pi/16$ x ½) radian from the ideal phase lead of $\pi/2$ [Mak 2003]. As it takes two points to calculate the momentum, the very first point (which is arbitrarily set to zero) should be ignored. (c) shows the down 4 momentum signal. It can be seen that the phase difference between the downsampled momentum and the downsampled price (in Fig 11.3(a)) is preserved. Again, the very first point (at $\omega = 0$) should be ignored.

(2) The data in the long-term frame contain aliases, which is a mixture of frequencies from the original signal in the intermediate frame. For example, after downsampling 5, the signal of the $\pi/2$ frequency in the altered signal can contain a mixture of signals of frequencies

$$\pi/10,\ \pi/10 + 2\pi/5,\ \pi/10 + 4\pi/5,\ \pi/10 + 6\pi/5,\ \pi/10 + 8\pi/5$$

from the original signal [Mak 2003].

EMA of the signal of the $\pi/2$ frequency in the long-term frame can thus contain EMA's of the mixture of signals of the above frequencies from the intermediate timeframe.

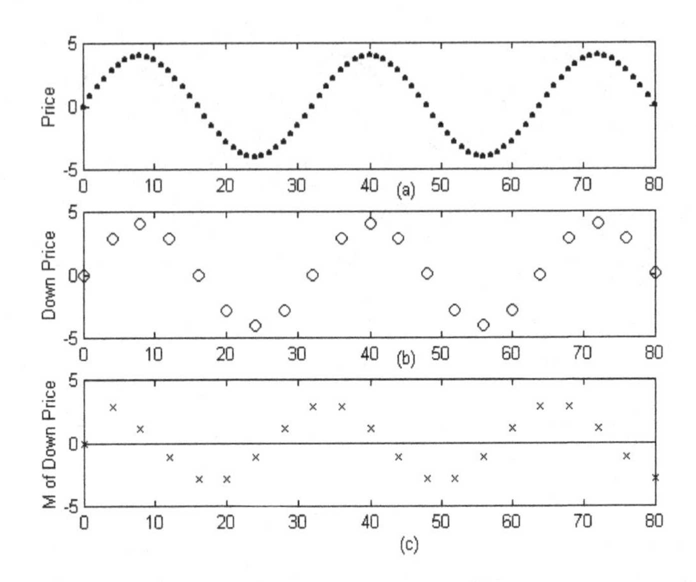

Fig 11.4(a) shows the original price data having a frequency of $\pi/16$ radian. (b) shows the down 4 price data. (c) shows the momentum indicator applied to the down 4 price data. As it takes two points to calculate the momentum, the very first point (which is arbitrarily set to zero) should be ignored.

(3) The EMA's of the mixtures of signals contain different amplitudes and phase lags or leads. Fig 11.5 plots the amplitude and phase of EMA3 (EMA of length 3) and EMA6 (EMA of length 6) from 0 to 5π. It can be noted that while the amplitude falls between 0 to π, it rises between π to 2π. Also, while there is a phase lag between 0 to π, there is a phase lead between π to 2π. Fig 11.6 plots a market price simulated by a pure sine wave of frequency $\pi/4$. Downsampling 5 would yield a signal of frequency $5\pi/4$. The EMA of length 3 of this signal is plotted in the figure (denoted as +) in the format of skipped convolution. This would represent how the EMA of the signal would look like in the long-term chart, but in a more detailed fashion [Mak 2003]. An EMA of length 13 (EMA13), which should yield similar smoothness of the original signal is also plotted (denoted as x) for comparison. In consistent with Fig 11.5, the EMA3 of signal of frequency $5\pi/4$ has an amplitude larger than the EMA13 of the original signal, and a phase lead compared to the original signal.

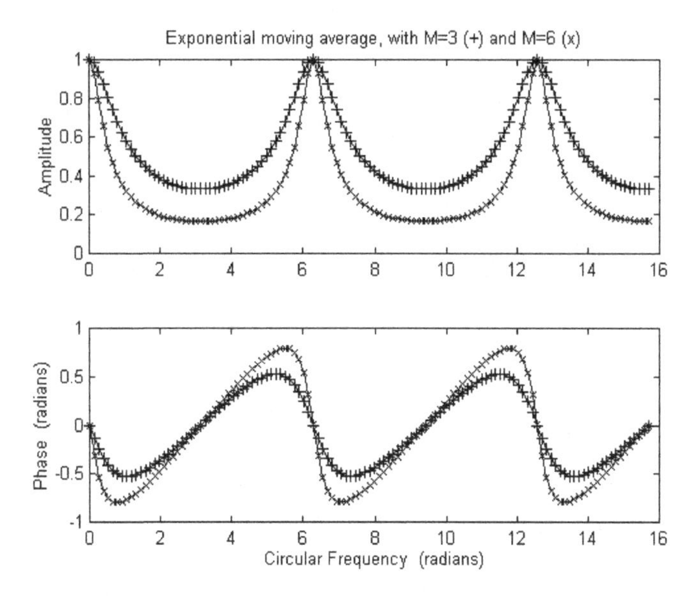

Fig 11.5 The top figure plots the amplitudes of the Fourier Transform of the exponential moving average of length 3 (plotted as +) and length 6 (plotted as x) versus the circular frequency ω from 0 to 5π. It can be noted that while the amplitude falls between 0 to π, it rises between π to 2π. The bottom figure plots the phases of the Fourier Transform of the exponential moving average of length 3 (plotted as +) and length 6 (plotted as x) versus the circular frequency ω from 0 to 5π. While there is a phase lag between 0 to π, there is a phase lead between π to 2π.

Thus, reducing the data from a timeframe to a timeframe with a longer term would have several disadvantages. Fortunately, with respect to the second and third disadvantages, the situation is not so bad, as higher frequencies (smaller cycle period) in market data has proportionately smaller amplitudes than the amplitudes in lower frequencies (longer cycle period). This amplitude variation has been pointed out by Ehlers [1992]. The amplitude of a market cycle has been observed empirically to be in proportion to the selected time scale. Thus, if the time and price scales were removed from the daily and weekly bar charts for the same commodity, one would not be able to tell which one is weekly, and which one is daily. This observation is consistent with the

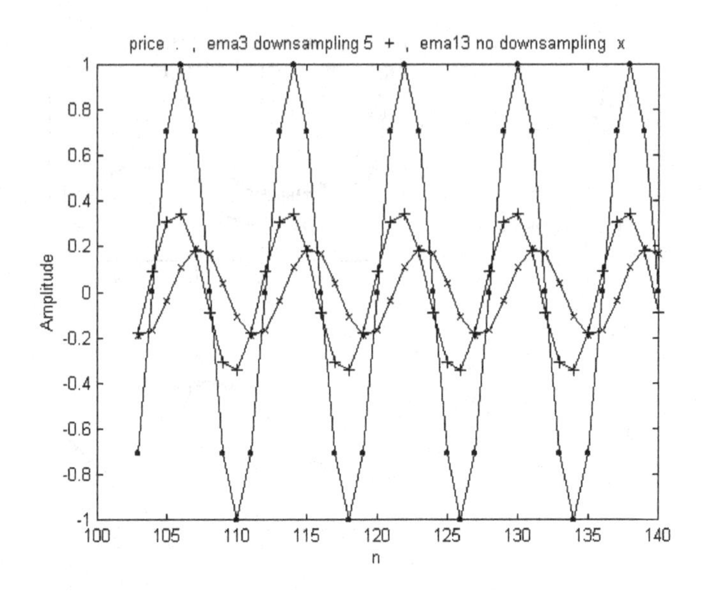

Fig 11.6 plots a market price simulated by a pure sine wave of frequency $\pi/4$ (plotted as . and joined by a line). Downsampling 5 would yield a signal of frequency $5\pi/4$. The EMA of length 3 (EMA3) of this signal is plotted in the figure (plotted as +) in the format of skipped convolution. This would represent how the EMA of the signal would look like in the long-term chart, but in a more detailed fashion [Mak 2003]. An EMA of length 13 (EMA13), which should yield similar smoothness of the original signal is also plotted (plotted as x) for comparison. In consistent with Fig 11.5, the EMA3 of signal of frequency $5\pi/4$ has an amplitude larger than the EMA13 of the original signal, and a phase lead compared to the original signal.

findings by Mantegna and Stanley [1995], who discovered that the successive variations of the S&P 500 index, Z, scaled with respect to the time interval, Δt. Scaling behavior was observed for time intervals spanning three orders of magnitude, from 1 minute to 1000 minutes. The probability distribution for the scaled variable, $Z/(\Delta t)^{0.712}$ is approximately the same for all these time intervals.

11.2 Multiple Screen Trading System

As mentioned earlier, a triple screen trading system has been suggested by Elder [1993, 2002]. We can look at the system in the following perspective. The first screen would preferably represent a low frequency that portraits the market trend. The second screen would represent a middle frequency that offers buying or selling opportunities. The third screen would represent a high frequency that triggers a buy or sell order. As the first two screens are used for decision making, and the third screen is used for placing orders, we will only consider the first two screens below.

The first screen is a long-term chart, which is taken to be weekly by Elder. A trend-following indicator should be applied to the data in the long-term chart, and a strategic decision should be made whether to trade long or short, or to stand aside. The original version of the Triple Screen Trading System employed the slope of the weekly MACD-Histogram as its weekly trend-following indicator [Elder 1993]. This, of course, is not correct. MACD-Histogram, as discussed earlier, is not a low-pass trending indicator, but a band-pass indicator. This indicator was eventually found to be very sensitive, yielding many buy and sell signals. It was then replaced by the slope of a weekly exponential moving average, which is used as the main trend-following indicator on long-term charts [Elder 2002]. When the weekly EMA rises, it identifies a bull move, and the trader should go long. When it falls, it indicates a bear move, and the trader should go short. As EMA is a trending indicator, it suits well for the first screen.

The second screen is an intermediate-term chart, which is taken to be daily by Elder. It was suggested that oscillators should be used to look for trading opportunities in the direction of the long-term trend. When the weekly trend is up, the trader should wait for the daily oscillators to fall, suggesting buy signals. When the weekly trend is down, the trader should look for daily oscillators to rise, giving sell signals. One of the choices of oscillators is the MACD-Histogram [Elder 2003]. This is not exactly correct. MACD-Histogram is not an oscillator, which is a high-pass filter. As we will show later, the slope of the MACD-Histogram should be used. The MACD-Histogram, acting as a band-pass filter, retains the middle frequency. The slope of the middle frequency would signify buying or selling opportunities. As a matter of fact, even though Elder [2002] plotted MACD-Histogram in his second screen, he was using the turning points of the MACD-Histogram to indicate trading opportunities. This is thus consistent with what we are suggesting here.

When the trend in the first screen is up, the trader should wait for the indicator in the second screen to go from negative to positive before executing a buy order. He can take profit when the indicator in the second screen goes from positive to negative. However, he should not sell short at this point.

When the trend in the first screen is down, the trader should wait for the indicator in the second screen to go from positive to negative before executing a sell order. He can take profit when the indicator in the second screen goes from negative to positive. However, he should not go long at this point.

We will devise here a trading plan, which is a slight modification of what is suggested by Elder [2002]:

(1) In the first screen, instead of the EMA of length 26 (used by Elder) applied to the market price in the long-term timeframe, we will use an EMA of length 130 in an intermediate timeframe. As discussed earlier, an EMA of length 130 has similar smoothness capability as an EMA of length 26 in a down 5 long-term timeframe. However, operating in the intermediate timeframe would eliminate some of the disadvantages of the long-term timeframe. A convolution of a velocity indicator is constructed on the EMA of length 130 and is plotted. If the velocity is positive, the trader will consider going long. If the velocity is negative, then the trader will consider going short. The cubic velocity indicator introduced by Mak [2003] is used here as it can easily tell whether the slope that it represents is positive or negative. It saves the effort of eyeballing the slope of the EMA, which is what Elder [2002] does.

(2) In the second screen, an intermediate-term timeframe would be used as suggested by Elder [2002]. However, instead of the MACD-Histogram used by Elder [2002], we will be using the cubic velocity of the MACD-Histogram. This would serve as an oscillator. The default values of the MACD-Histogram used by traders (12, 26 and 9) will be used. When the velocity in the first screen is positive, the trader would buy when the oscillator in the second screen goes from negative to positive. He will take profit when the

velocity in the second screen becomes negative. When the velocity in the first screen is negative, the trader would sell when the oscillator in the second screen goes from positive to negative. He will take profit when the velocity in the second screen becomes positive.

Fig 11.7(a) plots the amplitude responses of the EMA of length 130 and the MACD-Histogram. Fig 11.7(b) plots the phase responses of the EMA of length 130 and the MACD-Histogram. It can be seen from Fig 11.7(a) that the EMA of length 130 retains the low frequency of the signal and eliminates other frequencies, especially the high frequency, while the MACD-Histogram retains the middle frequency and attempts to eliminate the low frequency and the high frequency.

11.2.1 *Examples of a Trading System*

Theoretical waveforms with a low, middle and high frequencies will be used as examples here. Ideally, the EMA applied in the first screen will eliminate the middle and high frequencies and retain only the low frequency, and the MACD-Histogram applied in the second screen will eliminate the low and high frequencies and retain only the middle frequency.

(1) A price signal, p, of three frequencies is constructed as follows:

$$p = 4\sin(\pi n/100) + \sin(\pi n/20) + 0.1\sin(\pi n/4) \qquad (11.1)$$

where n is a positive integer.

This price signal is plotted in Fig 11.8(a). An exponential moving average (EMA) of length 130 (denoted by +) is applied to the price signal. A cubic velocity indicator is applied to the EMA and plotted in Fig 11.8(b), which forms the first screen. An MACD-Histogram of values 12, 26 and 9 is applied to the original price signal. The cubic velocity of the MACD-Histogram is then calculated and is plotted in Fig 11.8(c), which forms the second screen.

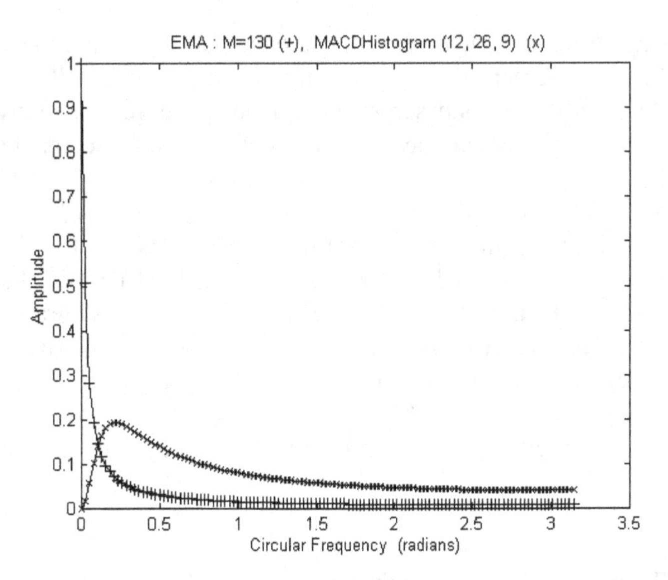

Fig 11.7(a) The amplitudes of the Fourier Transform of the EMA of length 130 (plotted as +) and the MACD-Histogram (plotted as x) are plotted versus circular frequency ω.

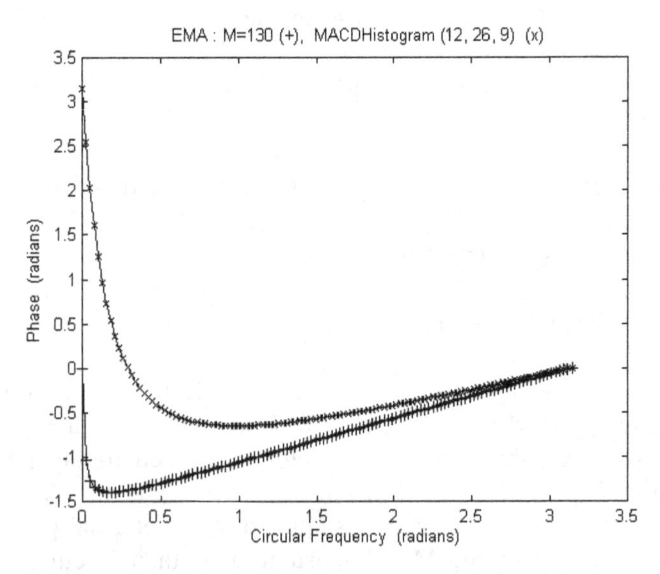

Fig 11.7(b) The phases of the Fourier Transform of the EMA of length 130 (plotted as +) and the MACD-Histogram (plotted as x) are plotted versus circular frequency ω.

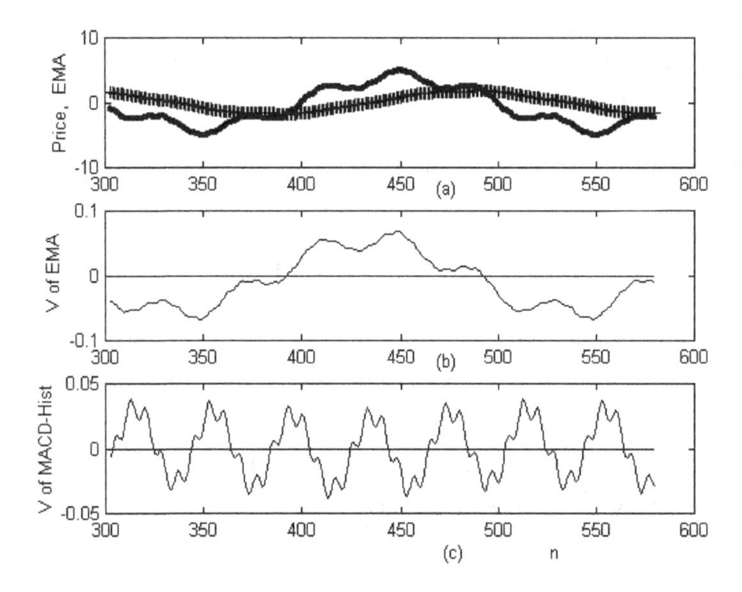

Fig 11.8(a) The price signal is plotted as a line. An exponential moving average (EMA) of length 130 (denoted by +) is applied to the price signal. (b) A cubic velocity indicator is applied to the EMA in (a). This forms the first screen in the trading system. (c) An MACD-Histogram of values 12, 26 and 9 is applied to the original price signal in (a). The cubic velocity of the MACD-Histogram is then calculated and is plotted. This forms the second screen in the trading system.

Ideally, the first screen should present a signal of the low frequency $\pi/100$, and the second screen should display a signal of the middle frequency $\pi/20$. This would provide traders with clear-cut buy and sell signals. In reality, as trending and oscillator indicators are never perfect, we would see a mixture of the two signals in both screens. Hopefully, the first screen would contain mostly the low frequency, and the second screen would contain mostly the middle frequency.

At n = 394, the velocity in the first screen (Fig 11.8(b)) changes from negative to positive. The trader should wait for the velocity in the second screen (Fig 11.8(c)) to go from negative to positive, which happens at n = 425. The trader should then buy and hold until n = 444, when the velocity in the second screen becomes negative.

At n = 494, the velocity in the first screen (Fig 11.8(b)) changes from positive to negative. The trader should wait for the velocity in the second screen (Fig 11.8(c)) to go from positive to negative, which happens at n = 525. The trader should then sell and hold until n = 544, when the velocity in the second screen becomes positive.

(2) A price signal, p, of three frequencies is constructed as follows:

$$p = 4\sin(\pi n/50) + \sin(\pi n/10) + 0.1\sin(\pi n/2) \tag{11.2}$$

where n is a positive integer.

This price signal is plotted in Fig 11.9(a). An exponential moving average (EMA) of length 130 (denoted by +) is applied to the price signal. A cubic velocity indicator is applied to the EMA and plotted in Fig 11.9(b), which forms the first screen. An MACD-Histogram of values 12, 26 and 9 is applied to the original price signal. The cubic velocity of the MACD-Histogram is then calculated and is plotted in Fig 11.9(c), which forms the second screen.

At n = 349, the velocity in the first screen (Fig 11.9(b)) changes from positive to negative. The trader should wait for the velocity in the second screen (Fig 11.9(c)) to go from positive to negative, which happens at n = 366. The trader should then sell and hold until n = 375, when the velocity in the second screen becomes positive.

At n = 399, the velocity in the first screen (Fig 11.9(b)) changes from negative to positive. The trader should wait for the velocity in the second screen (Fig 11.9(c)) to go from negative to positive, which happens at n = 416. The trader should then buy and hold until n = 425, when the velocity in the second screen becomes negative.

(3) The trading system, however, will not work for the above two examples if the amplitude of the high frequency signal is increased from 0.1 to 0.25.

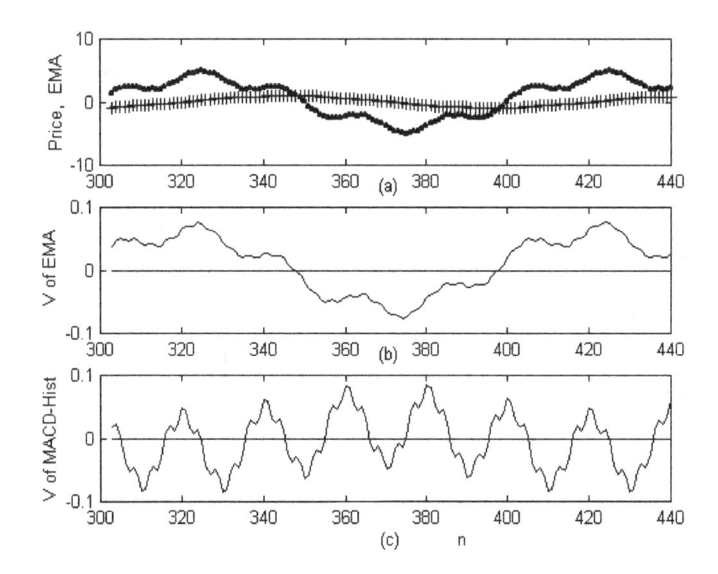

Fig 11.9(a) The price signal is plotted as a line. An exponential moving average (EMA) of length 130 (denoted by +) is applied to the price signal. (b) A cubic velocity indicator is applied to the EMA in (a). This forms the first screen in the trading system. (c) An MACD-Histogram of values 12, 26 and 9 is applied to the original price signal in (a). The cubic velocity of the MACD-Histogram is then calculated and is plotted. This forms the second screen in the trading system.

Take, for example, a price signal, p, of three frequencies constructed as follows:

$$p = 4\sin(\pi n/50) + \sin(\pi n/10) + 0.25\sin(\pi n/2) \tag{11.3}$$

where n is a positive integer.

This price signal is plotted in Fig 11.10(a). An exponential moving average (EMA) of length 130 (denoted by +) is applied to the price signal. A cubic velocity indicator is applied to the EMA and plotted in Fig 11.10(b), which forms the first screen. An MACD-Histogram of values 12, 26 and 9 is applied to the original price signal. The cubic velocity of the MACD-Histogram is then calculated and is plotted in Fig 11.10(c), which forms the second screen.

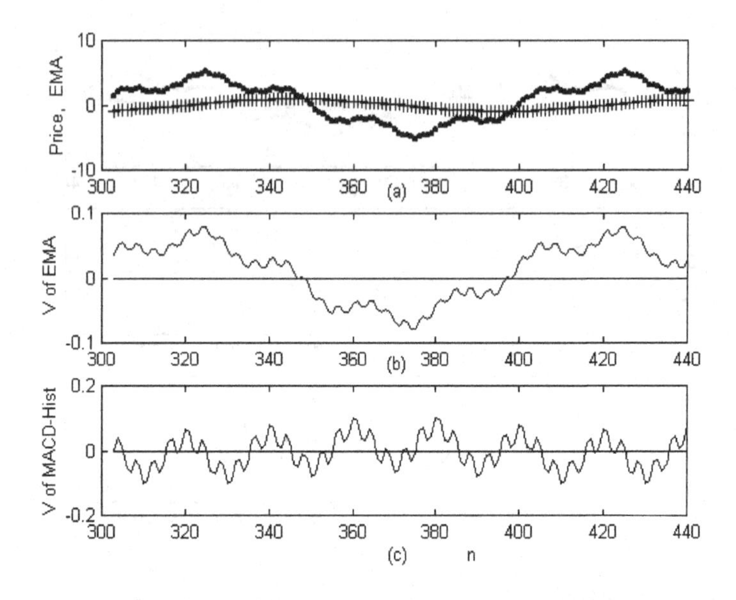

Fig 11.10(a) The price signal is plotted as a line. An exponential moving average (EMA) of length 130 (denoted by +) is applied to the price signal. (b) A cubic velocity indicator is applied to the EMA in (a). This forms the first screen in the trading system. (c) An MACD-Histogram of values 12, 26 and 9 is applied to the original price signal in (a). The cubic velocity of the MACD-Histogram is then calculated and is plotted. This forms the second screen in the trading system.

The first screen (Fig 11.10(b)) looks similar to but noisier than Fig 11.9(b). We can still easily tell when the market is trending up, and when it is trending down. However, in Fig 11.10(c), as MACD-Histogram does not eliminate high frequency signal well, the velocity of the MACD-Histogram, shown in this second screen (Fig 11.10(c)), is therefore, quite noisy. Thus, it is difficult to find buying and selling opportunities.

11.2.2 *Triple Screen Trading System*

In section 11.2.1, market price is modeled as a combination of a low frequency, a middle frequency, and a high frequency. The objective of the trading plan devised is to isolate the low frequency in the first screen and the middle frequency in the second screen. These two screens correspond to the first two screens of the Triple Screen suggested by Elder [1993, 2002]. The examples described in section 11.2.1 shows that the popular Triple Screen Trading System does make sense. However, its success would very much depend on the frequencies and amplitudes of the frequencies of the signal.

It is possible that the Triple Screen Trading System can be improved by using scaling function and wavelets. While the scaling function serves as a low pass filter, the wavelets can serve as band-pass filters [Mak 2003].

11.3 Test of a Trading System

As the frequencies (or periods) of a market price signal change quite often, there is only certain probabilities that a trading plan is profitable. A trader would not, of course, expect a trading plan to be profitable in all market conditions.

To test whether a trading plan is reasonable and profitable, the following steps should be performed:

(1) The frequency characteristics, i.e., the amplitude and phase, of the indicators applied to market data in the multiple screens should be studied.

(2) The trading plan should be applied to various theoretical waveforms of different frequencies and amplitudes to see what are its advantages and limitations.

(3) The trading plan should be applied to a large number of real market data to see whether it is profitable.

As any trading system would have only a limited probability of success in an ever-changing financial market, a trader should learn how to manage his money. We will deal with the money management issue in the next two chapters.

Chapter 12

Money Management — Time Independent Case

Management is the act of handling direction or control of a task. Ideally, one should use the minimum resources – time, effort, and money – to achieve the maximum success. The act of managing money - money management, is how one should utilize his money to attain the maximum reward.

In the financial market, a trader should realize that the task that he is performing has only a certain probability of success, and, as such, he should not put all his eggs in one basket. Furthermore, he needs to realize that he may run into a string of losses in a row. He thus needs to arrange his finance such that he can survive those draw-downs. A professional trader would arrange to cut his loss in every single trade. He would put a stop-loss order at the same time he puts in a buy or sell order. He usually would limit his loss to two percent of his equity in a single trade.

If a trader knows the probability distribution of the variations of share prices, it would definitely help him with his money management planning. The variation of share price is quite often taken as a random process. While this may not necessarily be true, it can be taken as a good approximation for calculation purpose. As described in Mantegna and Stanley [2000], idealization and approximation are quite common in scientific investigations. For example, in Physics, frictionless motion has been employed to develop laws in dynamics, while, in the real world, frictionless motion rarely happens. Thus, while the financial market may not be a random process, we may be able to devise some useful results assuming that it is.

12.1 Probability Distribution of Price Variation

We will take a look at the probability distribution of the price variation of the market and how money management technique can be designed with that information. The S & P 500 index would be taken as an example. The statistical distribution of the S & P 500 index has been studied for a time-scale that ranges from 1 minute to 1000 minutes during a six-year period from January 1984 to December 1989 [Mantegna and Stanley 1995]. It has been found that, except for the most rare events, it is a stochastic process that is quite well described by a Levy stable symmetrical process. The Levy distribution is given by:

$$L(Z, \Delta t) \equiv \frac{1}{\pi}\int_0^{\infty} \exp(-\gamma \Delta t q^{\alpha}) \cos(qZ) dq \qquad (12.1)$$

where

$Z = y(t + \Delta t) - y(t)$ is the successive index variation

$y(t)$ is the value of the S & P 500 index

Δt is taken to be 1 time unit

α is an index determined empirically to be 1.40 +/- 0.05

γ is a scale factor determined empirically as 0.00375

The notation, Z, has a different meaning from the same notation denoted in Section 2.1.3. It is used here to conform to the notation of Mantegna and Stanley [1995, 2000]. The probability distribution f(Z), which is given by Eq (12.1), is plotted in Fig 12.1. For Δt equals to 1, the standard deviation, σ, is determined from the experimental data and is equal to 0.0508 [Mantegna and Stanley 1995]. The experimental data agrees well with the theoretical Levy profile up to $|Z|/\sigma < 6$ (i.e., $|Z| < 0.3$ approximately). For $|Z|/\sigma > 6$, the data falls off exponentially from the stable distribution. The fall-off can be more easily visualized in Mantegna and Stanley [1995] where $\log_{10} f(Z)$ is plotted versus Z/σ.

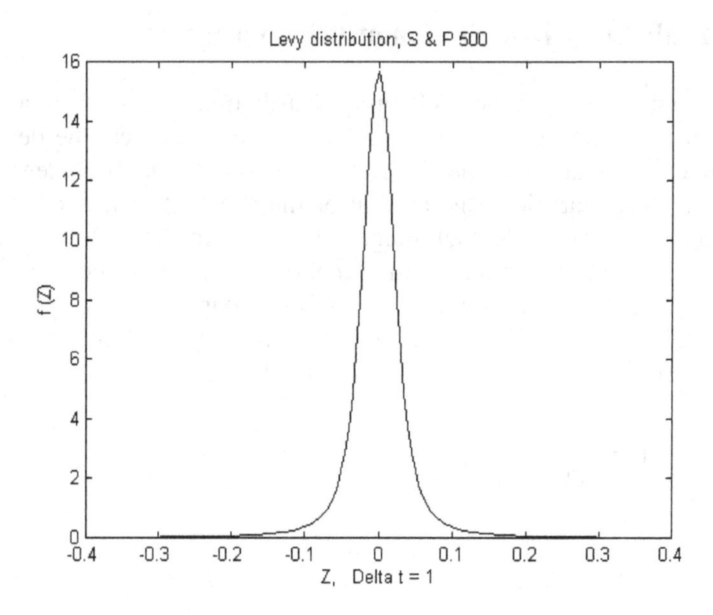

Fig 12.1 Levy stable symmetrical distribution of the S & P 500 for time interval $\Delta t = 1$ minute, $\alpha = 1.40$ and $\gamma = 0.00375$.

The S & P 500 data has been re-scaled for $\Delta t = 3$, 10, 32, 100, 316 and 1000 minutes under the transformations [Mantegna and Stanley 1995]

$$Z_m(\Delta t = 1) \equiv \frac{Z}{(\Delta t)^{1/\alpha}} \tag{12.2}$$

and

$$L_m(Z_m, \Delta t = 1) \equiv \frac{L(Z, \Delta t)}{(\Delta t)^{-1/\alpha}} \tag{12.3}$$

All the data collapse on the $\Delta t = 1$ minute distribution. Thus, it was concluded that a Levy distribution describes well the probability distribution of the S & P 500 index over time intervals spanning three orders of magnitude.

We have calculated the area under the curve $L(Z, \Delta t)$ for $\Delta t = 1$ (depicted in Fig 12.1) and obtained a result of 0.9996. We will consider the area equals to 1 for all calculations later. Thus, we will consider $L(Z, \Delta t) = f(Z)$ as a probability density function (pdf), as it satisfy the conditions [Freund 1992, Meyer 1966]:

(1) $f(Z) \geq 0 \quad \text{for} \quad -\infty < Z < \infty ;$ (12.4a)

(2) $\int_{-\infty}^{\infty} f(Z)dZ = 1$ (12.4b)

12.2 Probability of Being Stopped Out in a Trade

Given the probability density function $f(Z)$, we will be able to calculate the probability, p, that a trade will be stopped out if a trader sets his stop-loss value to be Z_{stop}. The probability, p, can be given by:

$$p = \int_{-\infty}^{s} f(Z)dZ \qquad (12.5)$$

where $s = Z_{stop} < 0$

We assume that the trader is buying and therefore going long. As the theory for selling short is very similar, the mathematical formulation would not be repeated here. Table 12.1 lists the Z_{stop} 's and their p values for $\Delta t = 1$. In the table, A is the area below the Levy distribution enclosed by Z_{stop} and $-Z_{stop}$ (which is positive). It represents the probability that the variation in Z will lie within Z_{stop} and $-Z_{stop}$. p can then be calculated as $(1 - A)/2$. When Z_{stop} is set at -0.05, which is approximately minus one standard deviation, A is approximately equal to 0.86. This can be compared with a standard normal distribution where A equals to 0.68 for one standard distribution. Thus, the Levy distribution is more advantageous to the trader than if the market had a normal distribution, as the probability of being stopped out is less. p' ($= 1 - p$) is the probability that the trade will not be stopped out if the stop-loss value is set to Z_{stop}. p is plotted versus Z_{stop} in Fig 12.2. It can be noted that p increases somewhat exponentially as Z_{stop}

approaches 0. The figure is significant as it shows the trader what kind of risk he would encounter. He needs to set his stop-loss such that his trade would not have a high probability of being stopped out.

Z_{stop}	A	p = (1-A)/2	p' = 1 - p	E(Z')
0	0	.5	.5	.0165
-0.0125	.3690	.3155	.6845	.0114
-0.025	.6319	.1841	.8159	.0083
-0.0375	.7805	.1097	.8903	.0066
-0.05	.8584	.0708	.9292	.0055
-0.0625	.9008	.0496	.9504	.0047
-0.075	.9259	.0371	.9629	.0042
-0.0875	.9419	.0291	.9709	.0038
-0.10	.9528	.0236	.9764	.0035

Table 12.1 The probability of being stopped out of a trade, p, and the expected value of a trade, E(Z') depends on the stop-loss value Z_{stop}. The calculation is performed for the S & P 500 index for time interval $\Delta t = 1$ minute. The index is assumed to be completely random.

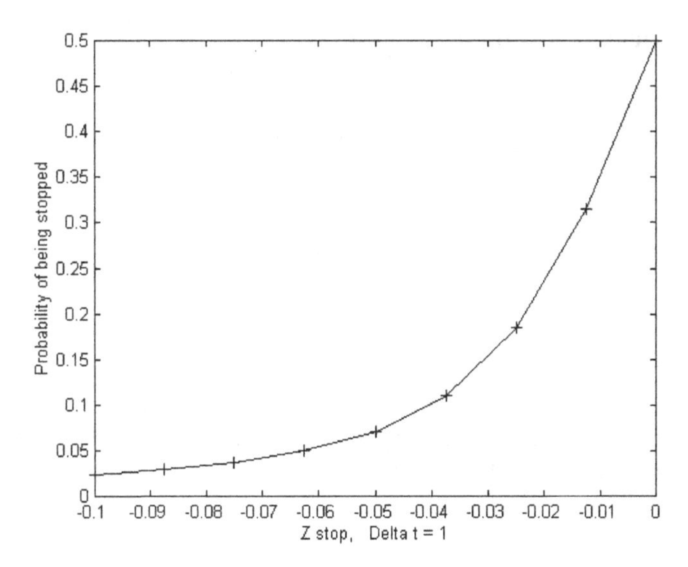

Fig 12.2 The probability of being stopped, p, at different Z_{stop} 's. The calculation is performed for the S & P 500 index for time interval $\Delta t = 1$ minute. The index is assumed to be completely random.

As f(Z) = L(Z), it can be shown from Eq (12.5), (12.2) and (12.3) that

$$p(Z, \Delta t) = p_m(Z_m, \Delta t = 1) \tag{12.6}$$

Thus, a trader trading in a timeframe with time unit Δt and setting his stoploss to Z, can find out the probability of his trade being stopped by using Eq (12.2) and then look it up in Table 12.1, where $Z_{stop} = Z_m$, and $p = p_m(Z_m, \Delta t = 1)$.

It should be noted that the data of Mantegna and Stanley [1995] employed here used closing prices of the market value, and did not consider high and low value during the time interval Δt. Thus, the actual probability of being stopped out should be higher. Furthermore, their data was taken from the market from January 1984 to December 1989. Traders who want to trade today's market need to check the current data and see whether any parameters have been changed from those listed in Eq (12.1)

12.3 Expected Value of a Trade

In probability theory, if Z' is a continuous random variable with probability density function (pdf) f, the expected value of Z' is defined as [Freund 1992, Meyer 1966]

$$E(Z') = \int_{-\infty}^{\infty} Zf(Z)dZ \tag{12.7}$$

where Z is a possible value of Z'. E(Z') is also referred to as the mean value of Z' . If f(Z) is symmetrical with respect to Z, the mean value is equal to 0. Therefore, if a trader enters a trade, and does not put a stop-loss order, his expected return is 0. However, if he puts in a stop-loss order, his expected return is given as follows:

$$E(Z') = s\int_{-\infty}^{s} f(Z)dZ + \int_{s}^{\infty} Zf(Z)dZ \tag{12.8}$$

where $s = Z_{stop}$ is the stop-loss value and is negative.

Eq (12.8) simply means that the trader limits his loss to the stop-loss value, s. If f(Z) is symmetrical with respect to Z, Eq (12.8) can be written as

$$E(Z') = \int_{-s}^{\infty} (Z+s)f(Z)dZ \tag{12.9}$$

As Z in the right-hand-side of Eq (12.9) is greater than or equal to –s (which is positive), the integrand inside the integral is greater than or equal to zero. Therefore, the expected value is always positive. It simply means that for a completely random market, the trader can make a profit in the long run if he puts stop-losses to his trades. Fig 12.3 plots the expected value of being stopped, E(Z'), at different Z_{stop} 's. The calculation is performed for the S & P 500 index for time interval $\Delta t = 1$ minute.

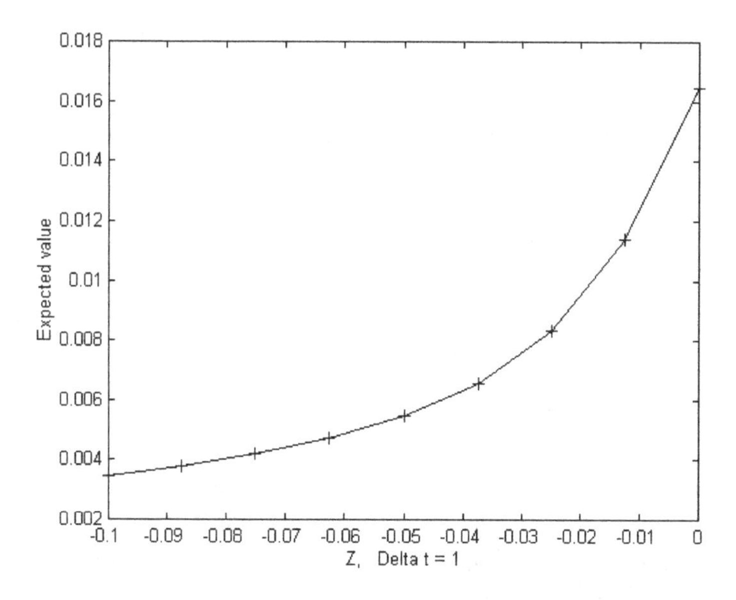

Fig 12.3 The expected value of being stopped, E(Z'), at different Z_{stop} 's. The calculation is performed for the S & P 500 index for time interval $\Delta t = 1$ minute. The index is assumed to be completely random.

If f(Z) is not symmetrical with respect to Z, Eq (12.8) can still be positive, depending upon how f(Z) is distributed. As long as Eq (12.8) is positive, the trader would have a positive gain in the long run.

Some professional traders claim that even if the market were random, with good money management, the market is still profitable. The result derived here supports their claim.

The expected values, E(Z'), of the S & P 500 is listed in Table 12.1 as a function of stop-loss values. As the S & P 500 index is discrete, but we take an approximation in our equations that it is continuous, the actual expected values should be slightly less than those listed in the Table. The expected value is largest when the stop-loss value is set equal to zero. However, no trader in his right mind would set the stop-loss value to be zero, as it corresponds to the largest probability where the trade will be stopped out. Thus, he has to find a compromise somewhere – a stop-loss value where he can afford the risk even in a string of losses (draw-down) but still will give him a reasonable profit in the long run.

The expected value can be re-scaled under the transformation:

$$E_m(Z_m', \Delta t = 1) \equiv \frac{E(Z', \Delta t)}{(\Delta t)^{1/\alpha}} \tag{12.10}$$

Thus, to find the expected value at certain time unit, Δt, the Z_{stop} and $E(Z')$ of each row in Table 12.1 should be multiplied by $(\Delta t)^{1/\alpha}$.

In this Chapter, we consider the trade being terminated in one time unit (e.g., 1 minute, 3 minutes, etc.). However, traders usually would hold the trade for much longer. What kind of gain or loss would he expect? We will consider this in the next Chapter. We will also show that, for the case of a trailing stop-loss, the scenario is consistent with the scenario described in this Chapter.

Chapter 13

Money Management — Time Dependent Case

In the last chapter, we discuss the scenario of a trader entering a trade and holding it for one time unit. If he puts in a stop-loss order every time he puts in a trade, he would expect a positive return in the long run.

However, a trader usually holds his trade for more than one time unit to increase his gain, if any. If so, at what time units (e.g., number of days) should he get out, in order to maximize his profit? Would that depend on the value of the stop-loss order he puts in? We will attempt to answer these questions in this Chapter. But first, we will describe some of the probability theory that needs to be used later.

13.1 Basic Probability Theory

13.1.1 *Experiment and the Sample Space*

In discussing probability, an experiment is defined as a process that generates well-defined outcomes or sample points. One and only one of the outcomes will occur. The set of all possible outcomes of the experiment forms the sample space, S. For example, in tossing a coin, the sample space is {head, tail}. In rolling a dice, the sample space is {1, 2, 3, 4, 5, 6}. When a trader puts a stop-loss order in his trade, the possible outcomes in day one will form a sample space {trade stopped, trade not stopped}.

13.1.2 *Events*

With respect to a particular sample space, an event is a set of possible outcomes, or a collection of sample points. In the terminology of set theory, an event is a subset of a sample space. For example, in the rolling of a dice, if we describe an event as when an odd number occurs, the event would be {1, 3, 5}.

Events can be combined to form new events. If A and B are events, the union of A and B, $A \cup B$ is the event which occurs if and only if A or B or both occur. Thus, the union of A and B contains all sample points belonging to A or B or both.

The intersection of A and B, $A \cap B$ is the event which occurs if and only if both A and B occur. Thus, the intersection of A and B contains the sample points belonging to both A and B.

Two events, A and B, are defined to be mutually exclusive if they cannot occur together. The two events have no sample points in common. This can be expressed as $A \cap B = \phi$; i.e., the intersection of A and B is an empty set.

Furthermore, the following properties are defined for the probability of A, P(A) [Meyer 1965]:

$$(1)\ 0 \le P(A) \le 1 \tag{13.1}$$

$$(2)\ P(S) = 1 \tag{13.2}$$

(3) If A and B are mutually exclusive events,

$$P(A \cup B) = P(A) + P(B) \tag{13.3}$$

If $A_1, A_2, \ldots..A_n, \ldots.$are pair-wise mutually exclusive events, it follows from property (3) that

$$P\left(\bigcup_{i=1}^{n} A_i\right) = \sum_{i=1}^{n} P(A_i) \tag{13.4}$$

Eq (13.4) is sometimes called the Addition Law for mutually exclusive events [Anderson et al 2005].

13.1.3 *Independent Events*

The probability of an event is quite often influenced by whether a related event has already occurred. The conditional probability of event A, given that event B has occurred is denoted by $P(A|B)$, which is read as "the probability of A given B". $P(A|B)$ is defined as [Meyer 1965]:

$$P(A \mid B) = \frac{P(A \cap B)}{P(B)} \qquad \text{provided that} \quad P(B) > 0 \tag{13.5}$$

There are many situations that the occurrence of event A has no bearing to whether event B has occurred. That is, events A and B are totally unrelated. The independence of two events A and B is defined as:

$$P(A \cap B) = P(A)P(B) \tag{13.6}$$

This definition is essentially equivalent to:

A and B are independent events if

$$P(A|B) = P(A) \tag{13.7a}$$

or

$$P(B|A) = P(B) \tag{13.7b}$$

Eq (13.6) is sometimes called the Multiplication Law for independent events [Anderson et al 2005]. Both the Addition Law for mutually exclusive events, Eq (13.4) and the Multiplication Law for independent events, Eq (13.6) will be used to calculate the probabilities of certain events in the financial market in the following sections.

13.2 Trailing Stop-Loss

A trailing stop-loss is a stop-loss where the trader moves the stop with respect to the increasing share price to protect his profit. The difference between the share price and the stop can be a constant or a function of the market price [Elder 1993, 2002] throughout the trade. We will only consider the difference being a constant here.

We will give an example. A trader buys at t = 0. The share price is arbitrarily set to x = 0. Any gain from then on is considered positive, and any loss negative. At t = 1, the trade can remain where it is, at x = 0, with a probability $p(x_0) = ½$, or increase to +1 with a probability $p(x_1) = ¼$, or decrease to –1 with a probability $p(x_{-1}) = ¼$ (Fig. 13.1(a)). If he puts a stop-loss order at x = -1, then he will be out of the market if the trade drops to –1 at t = 1. He will remain in the market only if the trade is at 0 or 1. If the trade is at x = 1, and he decides to put in a trailing

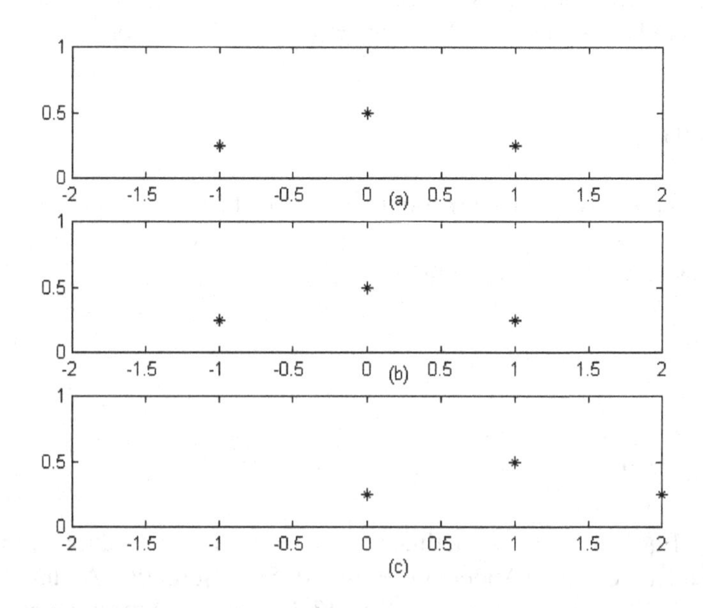

Fig 13.1(a) Probability distribution of the share price at t = 1. The original share price purchased at t = 0 was x = 0.
(b) and (c) Probability distribution of the share price at t = 2. For (b), the share price centers at 0 from t = 1. For (c), the share price centers at 1 from t = 1.

stop-loss order, he would advance his stop-loss from $x = -1$ to $x = 0$. The trade will then go onto the next time unit, i.e., $t = 2$. At $t = 2$, the trade can again, remain at $x = 0$, with a probability of ½, or increase to 1 with a probability of ¼, or decrease to –1 with a probability of ¼ (Fig 13.1(b)). If the trade decreases to –1, he will be out of the market. If the trade was at $x = 1$ at $t = 1$, it can remain at $x = 1$ with a probability of ½, or increase to $x = 2$ with a probability of ¼, or decrease to 0 with a probability of ¼ (Fig 13.1(c)). If the trade decreases to 0, it will be stopped out. If it increases to 2, the trader can again move the stop-loss from $x = 0$ to $x = 1$. As long as he is not being stopped, he can go onto the next time unit (e.g., the next day). Every time the trade advances, he will move the stop-loss order up. This would, hopefully increases his profit. Eventually, at $t = k$, he decides to cash out. We will now work out what is the probability that this will happen, and what is his expected gain, or loss.

13.2.1 *Probability and Expected Value*

Let S_i be the event that the trade will be stopped out of the market at $t = i$ and S_i' will be the event that the trade will not be stopped. Let C_k be the event that the trader decides to cash out the trade. Then the probability that the trade will not be stopped until the trader decides to cash out at $t = k$, will be given by:

$P_c(t = k)$

$= P(S_1' \cap S_2' \cap S_3' \ldots\ldots\ldots\ldots \cap S_{k-1}' \cap C_k)$

$= P(S_1')\, P(S_2')\, P(S_3') \ldots\ldots\ldots\ldots P(S_{k-1}')\, P(C_k)$

$$= P(S_1')\, P(S_2')\, P(S_3') \ldots\ldots\ldots\ldots P(S_{k-1}') \qquad (13.8)$$

as $P(C_k) = 1$

Thus the probability is the same as the probability that the trade has not been stopped at $t = k - 1$. In deriving Eq (13.8), the Multiplication Law for independent events, Eq (13.6), has been used, as the events are independent of each other.

Eq (13.8) can be written as

$$P_c(t=k) = \sum_{i_1=-n+1}^{m} p(x_{i_1}) \sum_{i_2=-n+1}^{m} p(x_{i_2}) R_{T_2,s} \cdots\cdots \sum_{i_{k-1}=-n+1}^{m} p(x_{i_{k-1}}) R_{T_{k-1},s} \qquad (13.9)$$

where

$p(x_i)$ is the probability that the trade will take on a value x_i ,
i = -m, ...-1, 0 , 1, ...m

$x_i = i\Delta d$, Δd is the interval between two adjacent x's.

$x_i < 0$ for $i < 0$

$x_0 = 0$

$x_i > 0$ for $i > 0$

x_{-n} ($= s < 0$) is the stop-loss value

The subscript j in i_j represents the jth time unit, where j = 1,2, ... k-1.

The lower bound of the summation sign is –n+1. This means that the sum starts from an x_i value that has not been stopped. It sums all the values that are larger than the stop-loss value.

T_j is one of the possible values that the trade has attained on the jth time unit (e.g. the jth day), and is given by

$$T_j = x_{i_1} + x_{i_2} + \ldots. + x_{i_{j-1}} + x_{i_j} = (i_1 + i_2 + \ldots\ldots.. + i_{j-1} + i_j)\Delta d \qquad (13.10)$$

$$R_{T_j,s} \begin{array}{ll} =1 & T_j > s \\ =0 & \text{otherwise} \end{array} \qquad (13.11)$$

R is a step function and is usually denoted by S [e.g., Butkov 1968]. It is denoted by R here so as not to be confused with the stop-loss value s. When T_j is larger than the original stop-loss value, s, R equals to 1. Otherwise, R equals to 0.

The expected value at t = k is given by:

$$E_c(t=k) = \sum_{i_1=-n+1}^{m} p(x_{i_1}) \sum_{i_2=-n+1}^{m} p(x_{i_2}) R_{T_2,s} \cdots\cdots \sum_{i_{k-1}=-n+1}^{m} p(x_{i_{k-1}}) R_{T_{k-1},s}$$

$$\times \sum_{i_k=-m}^{m} p(x_{i_k})[(1-D_r)s(1-D_t) + D_r(T_{k-1}+s)(1-D_t)$$

$$+ (1-D_s)sD_t + D_s T_k D_t] \qquad (13.12)$$

where

$$D_r = R_{T_{k-1}+s,s} \qquad (13.13a)$$

$$D_t = R_{T_k, T_{k-1}+s} \qquad (13.13b)$$

$$D_s = R_{T_k,s} \qquad (13.13c)$$

and

$$R_{a,b} = \begin{matrix} 1 & a > b \\ 0 & \text{otherwise} \end{matrix} \qquad (13.14)$$

While s represents the original stop-loss, $T_{k-1} + s$ represents the trailing stop-loss. $(1-D_r)$ is equal to 1 when the trailing stop-loss is less than or equal to the original stop-loss, or, simply, T_{k-1} is less than or equal to 0; otherwise, $(1-D_r)$ is equal to 0. $(1-D_t)$ is equal to 1 when T_k is less than or equal to the trailing stop-loss; otherwise, it is equal to 0. $(1-D_s)$ is equal to 1 when T_k is less than or equal to the original stop-loss; otherwise, it is equal to 0.

Thus, the first two terms inside the square bracket of Eq (13.12) means that T_k is less than or equal to the trailing stop-loss. However, the first term implies that the trade is stopped by the original stop-loss, and therefore takes on the value, s, while the second term implies that the trade is stopped by the trailing stop-loss, and therefore takes on the value of the trailing stop-loss. The third and fourth term inside the square bracket means that T_k is larger than the trailing stop-loss. However, the third term implies that the trade is stopped by the original stop-loss, as T_k is less than or equal to the stop-loss, and therefore takes on the stop-loss value, s. The fourth term implies that the trade is not stopped at all, but is cashed out by the trader, at whatever value, T_k , that the trade has attained.

Note that the last summation starts from –m, this means that for a trade that is being stopped, the sum adds up all the probability p(x) for x which is less than or equal to the stop-loss value, or the trailing stop-loss value.

As an example, we will choose the possible values of x_i to be (-0.075, -0.05, -0.025, 0, 0.025, 0.05, 0.075) and calculate the normalized probabilities at those values for the S & P 500 at Δt = 1 minute using Eq (12.1). This probability distribution is shown in Fig. 13.2. The probabilities have been normalized such that their sum is equal to 1.

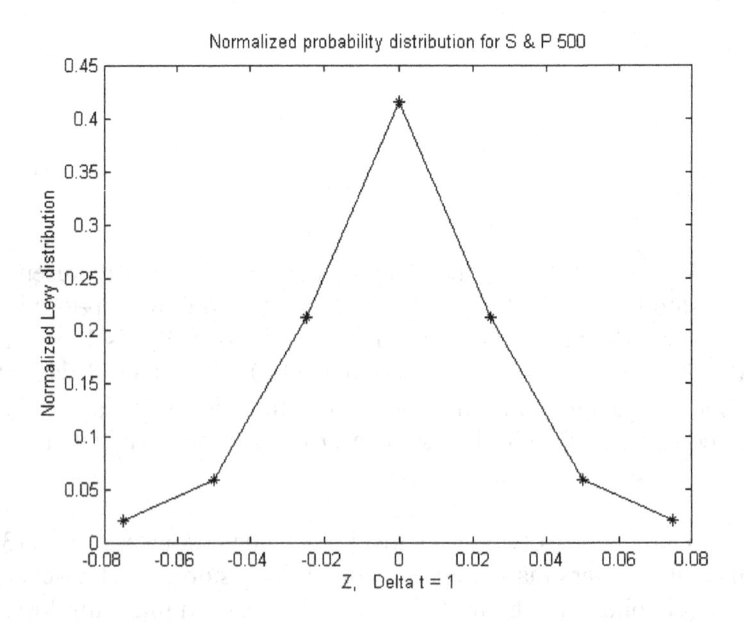

Fig 13.2 Example values of the normalized probabilities of the S & P 500 index at Δt = 1 minute, used for the calculation of the expected value of a trade. Here, Δd = 0.025.

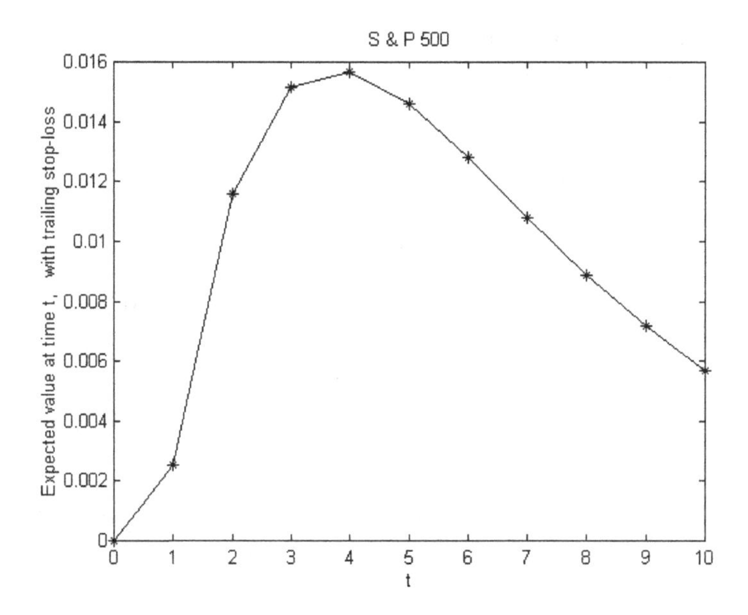

Fig 13.3 Expected value of a trade at t minutes, provided that the trade has not been stopped out of the market before t. The trailing stop-loss value is set equal to –0.025 S & P 500 index point.

Choosing the stop-loss value to be equal to –0.025, the expected value E_c is calculated and plotted versus time t, in Fig 13.3. It can be seen that E_c goes through a local maximum. The implication is that once the trader decides on what the stop-loss value is, he should cash out at t where E_c is maximum in order to gain maximum benefit. This, of course, assumes that the trade has not been stopped out before then.

13.2.2 *Total Probability and Total Expected Value*

In the last section, we discuss the probability and expected value of a trade at time t = k, assuming that the trade has not been stopped out before t = k. What is the probability and expected value if the trade has been stopped, either by the original stop-loss or the trailing stop-loss before then? And what are the total probability and the total expected value of all these events that can happen? We will attempt to answer these questions.

The total probability, P_T , that a trade may be stopped by the original stop-loss or the trailing stop-loss before t = k or the trader decides to cash the trade out at t = k, is given by:

$$P_T = P[S_1 \cup (S_1' \cap S_2) \cup (S_1' \cap S_2' \cap S_3) \cup \ldots$$

$$\ldots \cup (S_1' \cap S_2' \cap S_3' \ldots\ldots\ldots\ldots \cap S_{k-1})$$

$$\cup (S_1' \cap S_2' \cap S_3' \ldots\ldots\ldots\ldots \cap S_{k-1}' \cap C_k)]$$

$$= P(S_1) + P(S_1' \cap S_2) + P(S_1' \cap S_2' \cap S_3) + \ldots$$

$$\ldots P(S_1' \cap S_2' \cap S_3' \ldots\ldots\ldots\ldots \cap S_{k-1})$$

$$+ P(S_1' \cap S_2' \cap S_3' \ldots\ldots\ldots\ldots \cap S_{k-1}' \cap C_k)$$

$$= P_s(t=1) + P_s(t=2) + P_s(t=3) + \ldots. + P_s(t=k-1) + P_c(t=k)$$

(13.15)

S_1 means that the trade is being stopped at t = 1. $(S_1' \cap S_2)$ means that the trade is not stopped at t = 1, but is stopped at t = 2. $(S_1' \cap S_2' \cap S_3' \ldots\ldots\ldots\ldots \cap S_{k-1}' \cap C_k)$ means that the trade is not stopped before t = k, but is cashed out by the trader at t = k. In deriving Eq (13.15), the Summation Law for mutually exclusive events, Eq (13.4), has been used.

The probability that the trade is being stopped at t = j , $P_s(t = j)$ is given by

$$P_s(t=j) = \sum_{i_1=-n+1}^{m} p(x_{i_1}) \sum_{i_2=-n+1}^{m} p(x_{i_2}) R_{T_2,s} \ldots\ldots \sum_{i_{j-1}=-n+1}^{m} p(x_{i_{j-1}}) R_{T_{j-1},s}$$

$$\mathrm{x} \sum_{i_j=-m}^{m} p(x_{i_j})[(1-D_r)(1-D_t) + D_r(1-D_t) + (1-D_s)D_t]$$

(13.16)

where j = 1,2, … k-1

It can be shown that, given

$$\sum_{i=-m}^{m} p(x_i) = 1 \tag{13.17}$$

P_T is equal to 1.

This is simply because P_T in Eq (13.15) is equivalent to:

$$P_T = \sum_{i_1=-m}^{m} p(x_{i_1}) \sum_{i_2=-m}^{m} p(x_{i_2}) \ldots\ldots \sum_{i_{k-1}=-m}^{m} p(x_{i_{k-1}}) \sum_{i_k=-m}^{m} p(x_{i_k}) \tag{13.18}$$

As each summation is equal to 1, P_T is equal to 1.

The total expected value, E_T , that a trade may be stopped by the original stop-loss or the trailing stop-loss before t = k or the trader decides to cash the trade out at t = k, is given by:

$$\begin{aligned}
E_T = E[&S_1 \cup (S_1' \cap S_2) \cup (S_1' \cap S_2' \cap S_3) \cup \ldots \\
&\ldots \cup (S_1' \cap S_2' \cap S_3' \ldots\ldots\ldots\ldots \cap S_{k-1}) \\
&\cup (S_1' \cap S_2' \cap S_3' \ldots\ldots\ldots\ldots \cap S_{k-1}' \cap C_k)] \\
= E(&S_1) + E(S_1' \cap S_2) + E(S_1' \cap S_2' \cap S_3) + \ldots \\
&\ldots E(S_1' \cap S_2' \cap S_3' \ldots\ldots\ldots\ldots \cap S_{k-1}) \\
&+ E(S_1' \cap S_2' \cap S_3' \ldots\ldots\ldots\ldots \cap S_{k-1}' \cap C_k) \\
= E_s&(t = 1) + E_s(t = 2) + E_s(t=3) + \ldots + E_s(t = k-1) + E_c(t = k)
\end{aligned} \tag{13.19}$$

The expected value when the trade is being stopped at $t = j$, $E_s(t = j)$ is given by:

$$E_s(t=j) = \sum_{i_1=-n+1}^{m} p(x_{i_1}) \sum_{i_2=-n+1}^{m} p(x_{i_2}) R_{T_2,s} \cdots\cdots \sum_{i_{j-1}=-n+1}^{m} p(x_{i_{j-1}}) R_{T_{j-1},s}$$

$$\times \sum_{i_j=-m}^{m} p(x_{i_j})[(1-D_r)s(1-D_t) + D_r(T_{j-1}+s)(1-D_t)$$

$$+(1-D_s)sD_t] \qquad (13.20)$$

The total expected value, E_T, is plotted in Fig 13.4

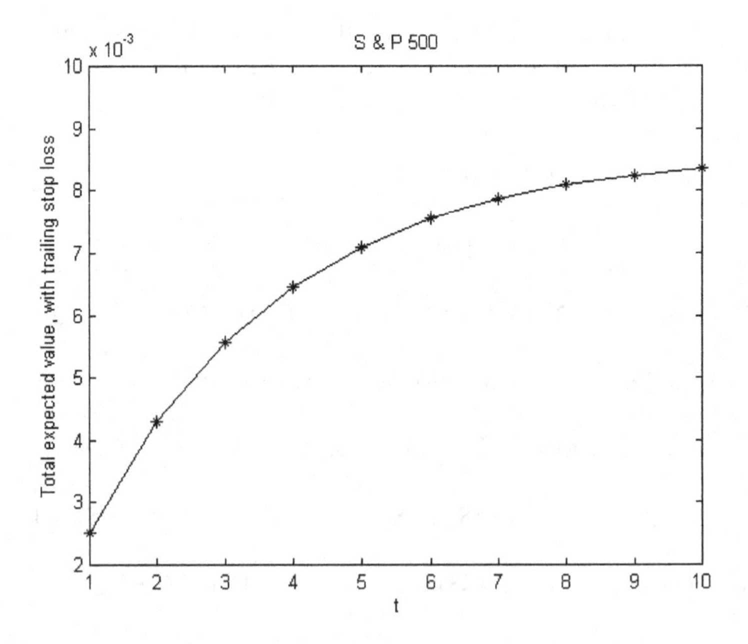

Fig 13.4 Total expected value of a trade at t minutes, when the trailing stop-loss value is set equal to –0.025 S & P 500 index point.

13.2.3 *Average Time*

As a trade can be stopped at t = 1, or t = 2, ..., or cashed out by the trader at t = k, the average time of a trade, E_t, can be described as the expected value of time that a trade is being stopped or cashed out, and is given by

$$E_t = P_s(t = 1) \times 1 + P_s(t = 2) \times 2 + P_s(t = 3) \times 3 + ... + P_s(t = k\text{-}1) \times (k\text{-}1) + P_c(t = k) \times k \quad (13.21)$$

13.2.4 *Total Expected Value/Average Time*

The average profit per average time unit, E_A , is given by:

$$E_A = E_T / E_t \quad (13.22)$$

Taking $s = x_{-1}$, E_A is found to be a constant with respect to t, whatever m is chosen to be (2m+1 is the number of possible discrete values of x_i , i.e, x_{-m} , x_{-m+1} , ..., x_{m-1} , x_m). Using the probability distribution depicted in Fig 13.2, E_A is calculated and plotted versus t, showing that it is a constant. This finding is surprising and interesting, especially considering that the numerator and denominator in Eq (13.22) has different mathematical formulations. The constant value, 0.00252, in Fig. 13.5 is much less than the value of 0.0083 listed in Table 13.1, as only a few x_i's are used in the calculation for illustration purpose (m is only taken to be 3 and Δd is taken to be 0.025). 0.0083 is considered a more accurate representation for a stop-loss value of –0.025. Taking s to be less than x_{-1} ,(e.g., x_{-2}), E_A will approach a near constant after the first few time units. Using a probability distribution depicted in Fig 13.6, and $s = x_{-2}$, E_A is calculated and plotted versus t in Fig 13.7. The initial lower value is caused by the trade being stopped by the original stop-loss. As t increases, most trades are stopped by the trailing stop-losses.

The implication of all these is that if the trader decides to stay in the market all the time (i.e., he will get back in the market as soon as he gets out), it does not matter which time unit (e.g., which day) he predetermines to cash out his trade. His profit would be the same. It can further be implied that he can leave his trade in the market, with a trailing stop-loss, and let the market stop him out. This, actually, is what a number of professional traders choose to do. This method has an advantage that if the market is not completely random, and is directional some of the time, the trader can rip the most benefit.

The above result can be compared with the result in the time-independent case. When a trader puts a trailing stop-loss to his trade, the difference between the market value and the stop-loss is always a constant. The result obtained, thus, should be, and is, consistent with that derived from the time-independent case.

The probability model employed here ignores commissions. Thus, the trade would be less profitable if the trader chooses to cash out his trade within the first few time units, providing he is not being stopped out before then.

A MATLAB program for calculating probabilities and expected value of a trade for the S & P 500 index using trailing stop-loss order is listed in Appendix 5. It can be modified to calculate other markets if the probability distribution of the other markets is used in the program.

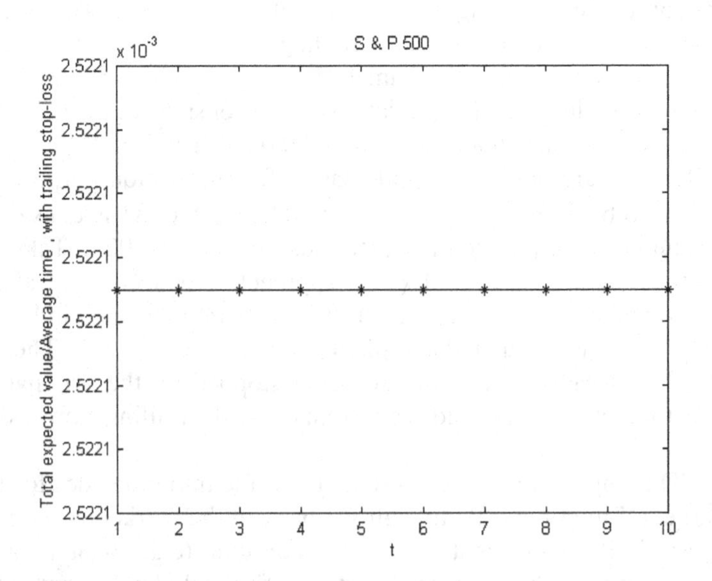

Fig 13.5 Total expected value/Average time of a trade at t minutes. The trailing stop-loss value is set equal to –0.025 (= x_{-1}) S & P 500 index point.

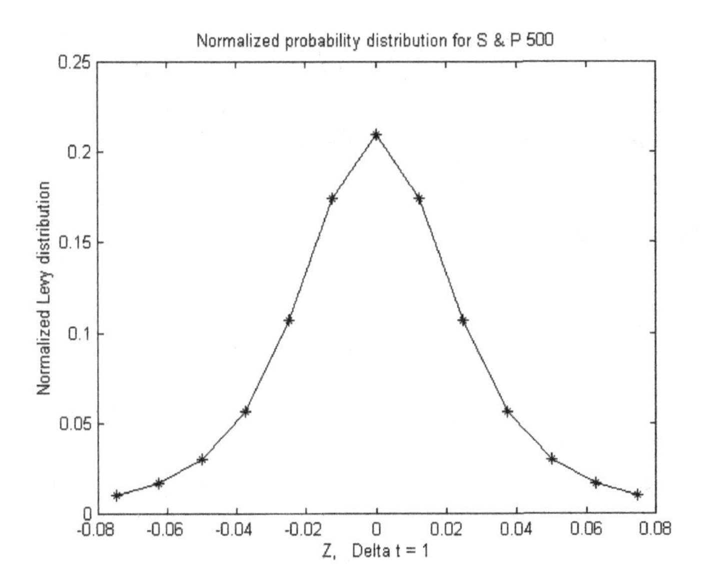

Fig 13.6 Example values of the normalized probabilities of the S & P 500 index at $\Delta t = 1$ minute, used for the calculation of the expected value of a trade. Here, $\Delta d = 0.0125$.

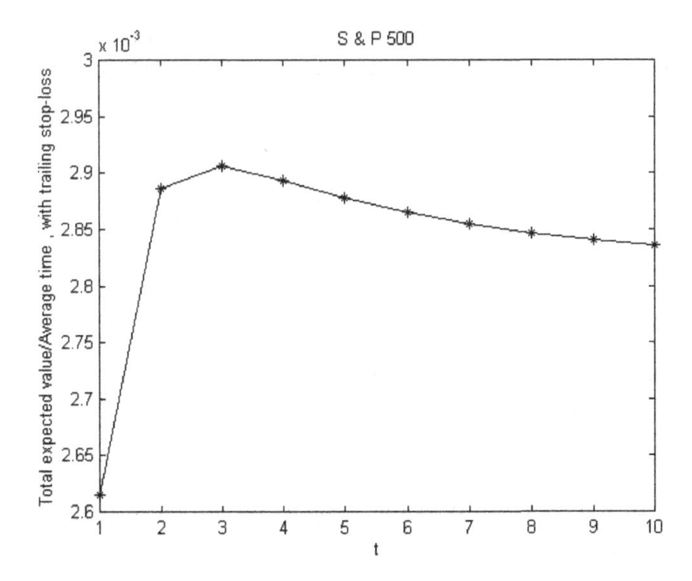

Fig 13.7 Total expected value/Average time of a trade at t minutes. The trailing stop-loss value is set equal to -0.025 ($= x_{-2}$) S & P 500 index point.

As a comparison to the trailing stop-loss technique, we will calculate the probabilities and expected values of a trade using a fixed stop-loss order in the next section. It will be shown that the fixed stop-loss technique is less profitable than the trailing one.

13.3 Fixed Stop-Loss

The trader can choose to fix his stop-loss instead of letting it trailing, i.e., he leaves the stop-loss value not changed during the whole trade. This technique has the advantage that he does not have to watch the market at all. He can decide to cash out his trade at t = k, provided that the trade has not been stopped out before then. We will now work out what is the probability that this can happen, and what is his expected gain, or loss.

13.3.1 *Probability and Expected Value*

Let S_i be the event that the trade will be stopped out of the market at t = i and S_i' will be the event that the trade will not be stopped. Let C_k be the event that the trader decides to cash out the trade. Then the probability that the trade will not be stopped until the trader decides to cash out at t = k, will be given by:

$P_c(t = k)$

$$= P(S_1' \cap S_2' \cap S_3' \ldots\ldots\ldots\ldots \cap S_{k-1}' \cap C_k)$$

$$= P(S_1')\, P(S_2')\, P(S_3') \ldots\ldots\ldots\ldots P(S_{k-1}')\, P(C_k)$$

$$= P(S_1')\, P(S_2')\, P(S_3') \ldots\ldots\ldots\ldots P(S_{k-1}') \qquad (13.23)$$

as $P(C_k) = 1$

Thus the probability is the same as the probability that the trade has not been stopped at t = k – 1.

Eq (13.23) can be written as

$$P_c(t=k) = \sum_{i_1=-m}^{m} p(x_{i_1})R_{T_1,s} \sum_{i_2=-m}^{m} p(x_{i_2})R_{T_2,s} \cdots\cdots \sum_{i_{k-1}=-m}^{m} p(x_{i_{k-1}})R_{T_{k-1},s} \qquad (13.24)$$

where

$p(x_i)$ is the probability that the trade will take on a value x_i ,

$i = -m, \ldots.-1, 0, 1, \ldots.m$

$x_i = i\Delta d$, Δd is the interval between two adjacent x's.

$x_i < 0$ for $i < 0$

$x_0 = 0$

$x_i > 0$ for $i > 0$

x_{-n} ($= s < 0$) is the stop-loss value

The subscript j in i_j represents the jth time unit, where j = 1, … k-1

T_j is one of the possible values that the trade has attained on the jth time unit (e.g. the jth day), and is given by

$$T_j = x_{i_1} + x_{i_2} + \ldots. + x_{i_{j-1}} + x_{i_j} = (i_1 + i_2 + \ldots\ldots.. + i_{j-1} + i_j)\Delta d \qquad (13.25)$$

$$R_{T_j,s} \begin{array}{ll} =1 & T_j > s \\ =0 & \text{otherwise} \end{array} \qquad (13.26)$$

When T_j is larger than the fixed stop-loss value, s, R equals to 1. Otherwise, R equals to 0.

The lower bound of the summation sign is –m, and the upper bound is m. However, the step function R, means that the sum starts from an x_i value that has not been stopped. It sums all the values that are larger than the stop-loss value.

The expected value at t = k is given by:

$$E_c(t=k) = \sum_{i_1=-m}^{m} p(x_{i_1})R_{T_1,s} \sum_{i_2=-m}^{m} p(x_{i_2})R_{T_2,s} \cdots \sum_{i_{k-1}=-m}^{m} p(x_{i_{k-1}})R_{T_{k-1},s}$$

$$\times \sum_{i_k=-m}^{m} p(x_{i_k})[s(1-R_{T_k,s}) + T_k R_{T_k,s}] \qquad (13.27)$$

The first term inside the square bracket means that the trade is being stopped by the fixed stop-loss value. The summation starts from –m, this means that for a trade that is being stopped, the sum adds up all the probability p(x) for x which is less than or equal to the fixed stop-loss value. The second term inside the square bracket means that the trader cashes out at whatever value the market has attained.

13.3.2 *Total Probability and Total Expected Value*

In the last section, we discuss the probability and expected value of a trade at time t = k, assuming that the trade has not been stopped out before t = k. What is the probability and expected value if the trade has been stopped by the fixed stop-loss before then? And what is the total probability and the total expected value of all these events that can happen?

The total probability, P_T, that a trade may be stopped by the fixed stop-loss before t = k or the trader decides to cash the trade out at t = k, is given by:

$$P_T = P[S_1 \cup (S_1' \cap S_2) \cup (S_1' \cap S_2' \cap S_3) \cup \ldots$$

$$\ldots \cup (S_1' \cap S_2' \cap S_3' \ldots\ldots\ldots\ldots \cap S_{k-1})$$

$$\cup (S_1' \cap S_2' \cap S_3' \ldots\ldots\ldots\ldots \cap S_{k-1}' \cap C_k)]$$

$$= P(S_1) + P(S_1' \cap S_2) + P(S_1' \cap S_2' \cap S_3) + \ldots$$

$$\ldots P(S_1' \cap S_2' \cap S_3' \ldots\ldots\ldots\ldots \cap S_{k-1})$$

$$+P(S_1' \cap S_2' \cap S_3' \ldots\ldots\ldots\ldots \cap S_{k-1}' \cap C_k)$$

$$= P_s(t = 1) + P_s(t = 2) + P_s(t=3) + \ldots. + P_s(t = k-1) + P_c(t = k) \tag{13.28}$$

S_1 means that the trade is being stopped at t = 1. $(S_1' \cap S_2)$ means that the trade is not stopped at t = 1, but is stopped at t = 2. $(S_1' \cap S_2' \cap S_3' \ldots\ldots\ldots\ldots \cap S_{k-1}' \cap C_k)$ means that the trade is not stopped before t = k, but is cashed out by the trader at t = k. In deriving Eq (13.28), the Summation Law for mutually exclusive events, Eq (13.4), has been used.

The probability that the trade is being stopped at t = j , $P_s(t = j)$ is given by

$$P_s(t = j) = \sum_{i_1=-m}^{m} p(x_{i_1})R_{T_1,s} \sum_{i_2=-m}^{m} p(x_{i_2})R_{T_2,s} \ldots\ldots \sum_{i_{j-1}=-m}^{m} p(x_{i_{j-1}})R_{T_{j-1},s}$$

$$\times \sum_{i_j=-m}^{m} p(x_{i_j})(1 - R_{T_j,s}) \tag{13.29}$$

It can be shown that, given

$$\sum_{i=-m}^{m} p(x_i) = 1 \tag{13.30}$$

P_T is equal to 1.

This is simply because P_T is equivalent to:

$$P_T = \sum_{i_1=-m}^{m} p(x_{i_1}) \sum_{i_2=-m}^{m} p(x_{i_2}) \ldots\ldots \sum_{i_{k-1}=-m}^{m} p(x_{i_{k-1}}) \sum_{i_k=-m}^{m} p(x_{i_k}) \tag{13.31}$$

As each summation is equal to 1, P_T is equal to 1.

The total expected value, E_T, that a trade may be stopped by the fixed stop-loss before t = k or the trader decides to cash the trade out at t = k, is given by:

$$E_T = E[S_1 \cup (S_1' \cap S_2) \cup (S_1' \cap S_2' \cap S_3) \cup \ldots$$

$$\ldots \cup (S_1' \cap S_2' \cap S_3' \ldots\ldots\ldots\ldots \cap S_{k-1})$$

$$\cup (S_1' \cap S_2' \cap S_3' \ldots\ldots\ldots\ldots \cap S_{k-1}' \cap C_k)]$$

$$= E(S_1) + E(S_1' \cap S_2) + E(S_1' \cap S_2' \cap S_3) + \ldots$$

$$\ldots E(S_1' \cap S_2' \cap S_3' \ldots\ldots\ldots\ldots \cap S_{k-1})$$

$$+ E(S_1' \cap S_2' \cap S_3' \ldots\ldots\ldots\ldots \cap S_{k-1}' \cap C_k)$$

$$= E_s(t = 1) + E_s(t = 2) + E_s(t = 3) + \ldots. + E_s(t = k\text{-}1) + E_c(t = k)$$

$$(13.32)$$

The expected value when the trade is being stopped at t = j , $E_s(t = j)$ is given by:

$$E_s(t = j) = \sum_{i_1=-m}^{m} p(x_{i_1}) R_{T_1,s} \sum_{i_2=-m}^{m} p(x_{i_2}) R_{T_2,s} \ldots\ldots \sum_{i_{j-1}=-m}^{m} p(x_{i_{j-1}}) R_{j_{k-1},s}$$

$$\text{x} \sum_{i_j=-m}^{m} p(x_{i_j}) s(1 - R_{T_j,s}) \qquad (13.33)$$

Using the probability distribution shown in Fig. 13.2 and choosing the fixed stop-loss value to be equal to –0.025, the total expected value, E_T, is plotted in Fig 13.8. Comparing the result with that plotted in Fig 13.4, it shows that a trade with a trailing stop-loss is more profitable than one with a fixed stop-loss.

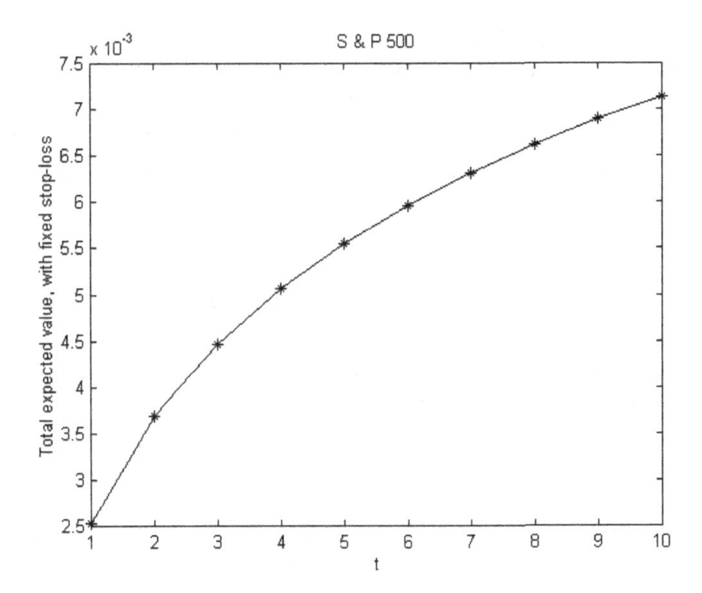

Fig 13.8 Total expected value of a trade at t minutes, when the fixed stop-loss value is set equal to –0.025 S & P 500 index point. This figure should be compared with Fig 13.4, where trailing stop-loss is used.

13.3.3 *Average Time*

As a trade can be stopped at $t = 1$, or $t = 2$, …, or cashed out by the trader at $t = k$, the average time of a trade, E_t, can be described as the expected value of time that a trade is being stopped or cashed out, and is given by

$$E_t = P_s(t = 1) \times 1 + P_s(t = 2) \times 2 + P_s(t = 3) \times 3 + \ldots + P_s(t = k-1) \times (k-1) + P_c(t = k) \times k \qquad (13.34)$$

13.3.4 *Total Expected Value/Average Time*

The average profit per average time unit, E_A, is given by:

$$E_A = E_T / E_t \qquad (13.35)$$

Using the probability distribution depicted in Fig 13.2, and taking $s = x_{-1}$, E_A is calculated and plotted versus t in Fig 13.9. It shows that E_A decreases from 0.00252 at t = 1 somewhat exponentially with respect to t. Thus, comparing Fig 13.9 with Fig 13.5, setting a trailing stop-loss is much more profitable than setting a fixed stop-loss for a trade. It is, therefore, no wonder, that the fixed stop-loss is not a money management technique preferred by traders.

A MATLAB program for calculating probabilities and expected value of a trade for the S & P 500 index using fixed stop-loss order is listed in Appendix 5. It can be modified to calculate other markets if the probability distribution of the other markets is used in the program.

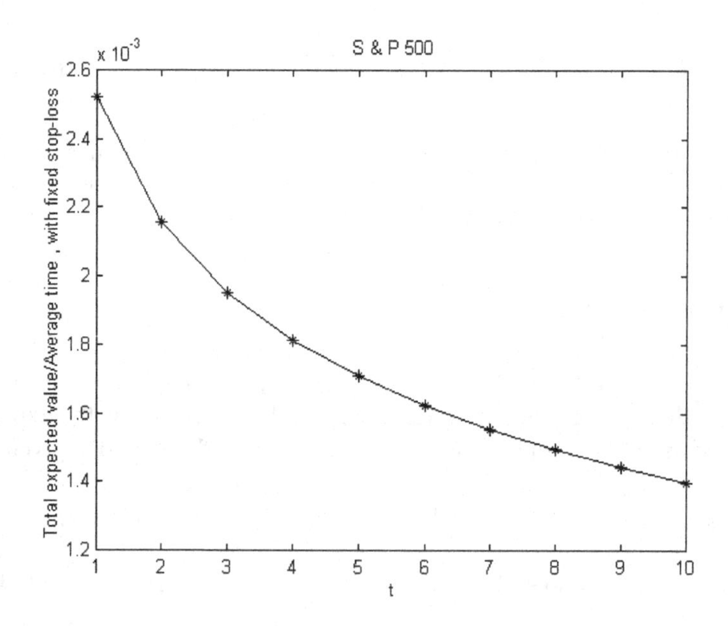

Fig 13.9 Total expected value/Average time of a trade at t minutes. The fixed stop-loss value is set equal to –0.025 S & P 500 index point. This figure should be compared with Fig 13.5, where a trailing stop-loss is used.

Chapter 14

The Reality of Trading

Trading the financial maket is not an easy task. It requires dedication, concentration and a lot of effort for the trader. Elder [2002] described the three M's of successful trading. The three M's are mind, method and money management.

14.1 Mind

Mind relates to the psychology of the trader. It includes discipline, record- keeping and training.

14.1.1 *Discipline*

The trader needs to take responsibility of his own action. He needs to set up certain rules and follow them. For example, he can move his stop only in the direction of his trade. If he starts to give more slack to his trade, he may wind up losing more money than he has expected.

14.1.2 *Record-Keeping*

A trader should keep good records. He needs good records to rate his performance and learn from his mistakes. He should keep a record of the chart of the market when he enters, with justification why he does so.

14.1.3 *Training*

Trading the market sucessfully needs years of training. The trader needs to spend his time studying and learning different strategies. He should prepare in advance what he would do when the market acts in various fashions. He should paper trade for a certain period before putting in real money. Elder [2002] compared the time and effort of the training of a trader to that of an airline pilot. This just shows the amount of dedication required to be a successful trader.

14.2 Method

Method would include fundamental analysis, technical analysis, and a hybrid of both [Mak 2003]. Fundamentalists look at prime rate, loan demand, price-to-earning ratio, etc., to detect why the price will change. Technical analysts ignore all these. They simply believe that past price and volume, and especially the former, tells all. All they need to do is to study the indicators that operate on the series of past financial data, and forecast which way the market is going. Traders who take a hybrid approach use both fundamental and technical analysis. They may use fundamentals to decide whether to buy or sell, and then use technical analysis to time the trades.

14.3 Money Management

Money management is especially important for beginners. Beginners should start out trading small sizes until he is profitable. The first priority is not to lose all the money. He can then work on making steady gains.

Traders, whether they are novices or professionals, should always put a stop-loss right after they put in a trade. They should not lose more than 2% of their equity in one trade.

Money management, together with a trader's mind and method, form the three components of successful trading. In this book, we have not discussed the psychology of a trader. There are many books on the subject. This book only talks about technical analysis and money management.

14.4 Technical Analysis

Indicators are tools of technical analysis. They analyse past market data and try to predict which way the market is heading. However, as is pointed out in Mak [2003], their properties are seldom understood. Their spectral contents have never been looked into. In order to fully understand an indicator, a spectrum analysis has to be performed. Their amplitude and phase response should be studied, with respect to frequency. The amplitude response would tell the trader what frequencies (or periods) the indicator has filtered out. This would show whether the indicator is relevant to the long-term, mid-term or short-term market movement. The phase response would tell the trader how fast the indicator is responding to the market movement, and how large a time lag he would expect. Some of the indicators have been analysed in this book. It has been shown that the significance of some indicators has been over-claimed by traders. However, some indicators, e.g., the MACD-Histogram, are very useful. It is interesting to note that the MACD-Histogram is actually a bandpass filter with very little lag. It is, therefore, no wonder that it is being used in a popular trading system called the Triple Screen trading system [Elder 2002].

Every trading system should be investigated by using spectrum analysis. A trading system is a strategy that a trader chooses to enter and exit his trade. It can consist of several screens to show the long and short term movements of the market. Understanding the indicator response in each screen would denote whether a trading system makes sense or not.

14.5 Probability Theory and Money Management

A trader should understand that he cannot win all the time. The market is somewhat random in nature. That is why it is very important that he puts a stop-loss to every trade to preserve his capital in case he loses. Some professional traders claim that even if the market were random, they can still make a profit with good money management. Using probability theory, we have shown that there is some truth in it. We have also shown that the trailing stop-loss technique practised by some professional traders is more profitable than the technique where the stop-loss is fixed during the whole trade.

Some professional traders, by their years of experience, do fine tune their tools, and sharpen their methods. They thus increase their probability of success. However, the tools and methodology should be understood by using mathematical analysis. This would allow new tools and new trading systems to be invented in the future. As the market is complex, traders need to stay ahead of the game all the time.

Appendix 1

Sinc Functions

A1.1 Coefficients of the Sinc Function with n = 2

The coefficients of the sinc function with n = 2, $h_2(k)$, are listed below with k = 0, 1, 2, …., 120.

h_2 =

0.5000	0.3183	0.0000	-0.1061	0.0000	0.0637	0.0000
-0.0455	0.0000	0.0354	0.0000	-0.0289	0.0000	0.0245
0.0000	-0.0212	0.0000	0.0187	0.0000	-0.0168	0.0000
0.0152	0.0000	-0.0138	0.0000	0.0127	0.0000	-0.0118
0.0000	0.0110	0.0000	-0.0103	0.0000	0.0096	0.0000
-0.0091	0.0000	0.0086	0.0000	-0.0082	0.0000	0.0078
0.0000	-0.0074	0.0000	0.0071	0.0000	-0.0068	0.0000
0.0065	0.0000	-0.0062	0.0000	0.0060	0.0000	-0.0058
0.0000	0.0056	0.0000	-0.0054	0.0000	0.0052	0.0000
-0.0051	0.0000	0.0049	0.0000	-0.0048	0.0000	0.0046
0.0000	-0.0045	0.0000	0.0044	0.0000	-0.0042	0.0000
0.0041	0.0000	-0.0040	0.0000	0.0039	0.0000	-0.0038

0.0000 0.0037 0.0000 -0.0037 0.0000 0.0036 0.0000

-0.0035 0.0000 0.0034 0.0000 -0.0034 0.0000 0.0033

0.0000 -0.0032 0.0000 0.0032 0.0000 -0.0031 0.0000

0.0030 0.0000 -0.0030 0.0000 0.0029 0.0000 -0.0029

0.0000 0.0028 0.0000 -0.0028 0.0000 0.0027 0.0000

-0.0027 0.0000

A1.2 Coefficients of the Sinc Function with n = 4

The coefficients of the sinc function with n = 4, $h_4(k)$, are listed below with k = 0, 1, 2, …., 120.

h_4 =

0.2500 0.2251 0.1592 0.0750 0.0000 -0.0450 -0.0531

-0.0322 0.0000 0.0250 0.0318 0.0205 0.0000 -0.0173

-0.0227 -0.0150 0.0000 0.0132 0.0177 0.0118 0.0000

-0.0107 -0.0145 -0.0098 0.0000 0.0090 0.0122 0.0083

0.0000 -0.0078 -0.0106 -0.0073 0.0000 0.0068 0.0094

0.0064 0.0000 -0.0061 -0.0084 -0.0058 0.0000 0.0055

0.0076 0.0052 0.0000 -0.0050 -0.0069 -0.0048 0.0000

0.0046 0.0064 0.0044 0.0000 -0.0042 -0.0059 -0.0041

0.0000 0.0039 0.0055 0.0038 0.0000 -0.0037 -0.0051

-0.0036 0.0000 0.0035 0.0048 0.0034 0.0000 -0.0033

-0.0045 -0.0032 0.0000 0.0031 0.0043 0.0030 0.0000

-0.0029 -0.0041 -0.0028 0.0000 0.0028 0.0039 0.0027

0.0000 -0.0026 -0.0037 -0.0026 0.0000 0.0025 0.0035

0.0025 0.0000 -0.0024 -0.0034 -0.0024 0.0000 0.0023

0.0032 0.0023 0.0000 -0.0022 -0.0031 -0.0022 0.0000

0.0021 0.0030 0.0021 0.0000 -0.0021 -0.0029 -0.0020

0.0000 0.0020 0.0028 0.0020 0.0000 -0.0019 -0.0027

-0.0019 0.0000

Appendix 2

Modified Low Pass Filters

Low pass filter removes high frequency components of a signal and allows low frequency components to pass.

A2.1 "Zero-lag" Exponential Moving Average

The output y(n) of the "zero-lag" exponential moving average [Ehlers 2001] can be written as

$$y(n) = \alpha\{x(n) + K[(x(n) - x(n-3)]\} + (1 - \alpha)y(n-1) \qquad \text{(A2.1)}$$

where

n is an integer and K = 0.5,

x(n) is the input price data.

Eq (A2.1) can be rewritten as

$$y(n) - (1-\alpha)y(n-1) = \alpha\{x(n) + K[(x(n) - x(n-3)]\} \qquad \text{(A2.2)}$$

The z-transform [Broesch 1997] of Eq (A2.1) is

$$Y(z) - (1-\alpha) z^{-1}Y(z) = \alpha [1 + K(1-z^{-3})] X(z) \qquad \text{(A2.3)}$$

where

$z = e^{i\omega}$

Y(z) is the transform of the output,

X(z) is the transform of the input.

The response function H(z) will be given by

$$H(z) = \frac{Y(z)}{X(z)} = \frac{\alpha[1 + K(1 - z^{-3})]}{1 - (1 - \alpha)z^{-1}} \qquad \text{(A2.4)}$$

By iterating the previously processed y value in Eq (A2.1), we can write y(n) as

$$y(n) = (1+K)\,\alpha\, x(n) + (1+K)\,\alpha\,(1-\alpha)\, x(n-1) + (1+K)\,\alpha\,(1-\alpha)^2\, x(n-2)$$
$$+ \sum_{j=3}^{\infty}[(1 + K)\alpha(1 - \alpha)^{j} - K\alpha(1 - \alpha)^{j-3}]\; x(n - j) \qquad \text{(A2.5)}$$

Thus, the first three coefficients are

$h(0) = (1+K)\,\alpha$

$h(1) = (1+K)\,\alpha\,(1-\alpha)$

$h(2) = (1+K)\,\alpha\,(1-\alpha)^2$ (A2.6)

The rest of the coefficients are given by

$h(j) = (1+K)\,\alpha\,(1-\alpha)^{j} - K\,\alpha\,(1-\alpha)^{j-3} \qquad j \geq 3$ (A2.7)

The two terms on the right hand side of Eq (A2.7) somewhat cancel each other, making the coefficients much less than the first three coefficients listed in Eq (A2.6).

A2.2 Modified EMA (MEMA), with a Skip 1 Cubic Velocity

The output y(n) can be written as

$$y(n) = \alpha \{ x(n) + K [11x(n)/6 - 3x(n-1) + 3x(n-2)/2 - x(n-3)/3] \} + (1-\alpha)\, y(n-1) \qquad \text{(A2.8)}$$

where

n is an integer,

x(n) is the input price data,

K = 1.

The response function H(z) will be given by

$$H(z) = \frac{\alpha[1 + K(11/6 - 3z^{-1} + (3/2)z^{-2} - (1/3)z^{-3})]}{1 - (1-\alpha)z^{-1}} \qquad \text{(A2.9)}$$

By iterating the previously processed y value in Eq (A2.8), we can write y(n) as

$$\begin{aligned} y(n) = {} & (1+11K/6)\,\alpha\, x(n) + [(1+11K/6)\,\alpha\,(1-\alpha) - 3K\alpha]x(n-1) \\ & + [(1+11K/6)\,\alpha\,(1-\alpha)^2 - 3K\alpha(1-\alpha) + 3K\alpha/2]x(n-2) \\ & + \alpha \sum_{j=3}^{\infty} [(1+11K/6)(1-\alpha)^j - 3K(1-\alpha)^{j-1} \\ & + (3K/2)(1-\alpha)^{j-2} - (K/3)(1-\alpha)^{j-3}]\; x(n-j) \qquad \text{(A2.10)} \end{aligned}$$

A2.3 Modified EMA (MEMA), with a Skip 2 Cubic Velocity

The output y(n) can be written as

$$y(n) = \alpha \{ x(n) + K [11x(n)/6 - 3x(n-2) + 3x(n-4)/2 - x(n-6)/3] \} + (1-\alpha)\, y(n-1) \qquad \text{(A2.11)}$$

where

n is an integer,

x(n) is the input price data,

K = 1/2.

The response function H(z) will be given by

$$H(z) = \frac{\alpha[1 + K(11/6 - 3z^{-2} + (3/2)z^{-4} - (1/3)z^{-6})]}{1 - (1-\alpha)z^{-1}} \qquad \text{(A2.12)}$$

By iterating the previously processed y value in Eq (A2.11), we can write y(n) as

$$\begin{aligned} y(n) = {} & (1+11K/6)\,\alpha\, x(n) + (1+11K/6)\,\alpha\,(1-\alpha)\, x(n-1) \\ & + [(1+11K/6)\alpha(1-\alpha)^2 - 3K\alpha\,]x(n-2) \\ & + [(1+11K/6)\,\alpha\,(1-\alpha)^3 - 3K\alpha(1-\alpha)]x(n-3) \\ & + [(1+11K/6)\alpha(1-\alpha)^4 - 3K\alpha(1-\alpha)^2 + 3K\alpha/2]x(n-4) \\ & + [(1+11K/6)\alpha(1-\alpha)^5 - 3K\alpha(1-\alpha)^3 + (3K\alpha/2)(1-\alpha)]x(n-5) \\ & + \alpha \sum_{j=6}^{\infty}[(1+11K/6)(1-\alpha)^j - 3K(1-\alpha)^{j-2} \\ & + (3K/2)(1-\alpha)^{j-4} - (K/3)(1-\alpha)^{j-6}]\ x(n-j) \end{aligned} \qquad \text{(A2.13)}$$

A2.4 Modified EMA (MEMA), with a Skip 3 Cubic Velocity

The output y(n) can be written as

$$y(n) = \alpha \{ x(n) + K [11x(n)/6 - 3x(n\text{-}3) + 3x(n\text{-}6)/2 - x(n\text{-}9)/3] \} + (1\text{-} \alpha) y(n\text{-}1) \qquad \text{(A2.14)}$$

where

n is an integer,

x(n) is the input price data,

K = 1/3.

The response function H(z) will be given by

$$H(z) = \frac{\alpha[1 + K(11/6 - 3z^{-3} + (3/2)z^{-6} - (1/3)z^{-9})]}{1 - (1-\alpha)z^{-1}} \qquad \text{(A2.15)}$$

By iterating the previously processed y value in Eq (A2.14), we can write y(n) as

$$\begin{aligned} y(n) = {} & (1+11K/6)\, \alpha\, x(n) + (1+11K/6)\, \alpha\, (1\text{-}\alpha)\, x(n\text{-}1) \\ & + (1+11K/6)\alpha(1\text{-}\alpha)^2\, x(n\text{-}2) \\ & + [(1+11K/6)\, \alpha\, (1\text{-}\alpha)^3 - 3K\alpha]x(n\text{-}3) \\ & + [(1+11K/6)\alpha(1\text{-}\alpha)^4 - 3K\alpha(1\text{-}\alpha)]x(n\text{-}4) \\ & + [(1+11K/6)\alpha(1\text{-}\alpha)^5 - 3K\alpha(1\text{-}\alpha)^2]x(n\text{-}5) \\ & + [(1+11K/6)\alpha(1\text{-}\alpha)^6 - 3K(1\text{-}\alpha)^3 + 3K\alpha/2]x(n\text{-}6) \\ & + [(1+11K/6)\alpha(1\text{-}\alpha)^7 - 3K(1\text{-}\alpha)^4 + (3K\alpha/2)(1\text{-}\alpha)]x(n\text{-}7) \end{aligned}$$

$$+[(1+11K/6)\alpha(1-\alpha)^8 - 3K(1-\alpha)^5 + (3K\alpha/2)(1-\alpha)^2]x(n-8)$$

$$+\alpha \sum_{j=9}^{\infty}[(1+11K/6)(1-\alpha)^j - 3K(1-\alpha)^{j-2}$$

$$+(3K/2)(1-\alpha)^{j-6} - (K/3)(1-\alpha)^{j-9}] \;\; x(n-j) \qquad \text{(A2.16)}$$

Appendix 3

Frequency

A3.1 Derivation of Frequency (4 points)

The financial market price data will be modeled as a sine wave superimposed on a constant level:

$$x = A \sin(\omega t + \phi) + D \quad \text{(A3.1)}$$

where x is the market price,
A is the amplitude of the sine wave,
ω is the circular frequency of the sine wave,
ϕ is the phase when time $t = 0$,
D is the constant level.

The prices x's would have been given. In order to solve for the four unknowns, A, ω, ϕ and D, four equations are required. From Eqn (A3.1), the four equations are chosen as follows:

$$t = 0 \quad x_0 = A \sin \phi + D \quad \text{(A3.2)}$$
$$t = -1 \quad x_{-1} = A \sin(-\omega + \phi) + D \quad \text{(A3.3)}$$
$$t = -2 \quad x_{-2} = A \sin(-2\omega + \phi) + D \quad \text{(A3.4)}$$
$$t = -3 \quad x_{-3} = A \sin(-3\omega + \phi) + D \quad \text{(A3.5)}$$

where x_0 is the closing price of the current bar,
x_{-1} is the closing price of one bar ago,
x_{-2} is the closing price of two bars ago,
x_{-3} is the closing price of three bars ago,

Subtracting Eqn (A3.5) from Eqn (A3.2), we get

$$x_0 - x_{-3} = A\,[2\cos(-3\omega/2 + \phi)\sin(3\omega/2)] \qquad \text{(A3.6)}$$

Subtracting Eqn (A3.4) from Eqn (A3.3), we get

$$x_{-1} - x_{-2} = A\,[2\cos(-3\omega/2 + \phi)\sin(\omega/2)] \qquad \text{(A3.7)}$$

Dividing Eqn (A3.6) by Eqn (A3.7) will yield

$$\frac{x_0 - x_{-3}}{x_{-1} - x_{-2}} = \frac{\sin(3\omega/2)}{\sin(\omega/2)} = 3 - 4\sin^2(\omega/2) \qquad \text{(A3.8)}$$

Therefore, the circular frequency, ω, can be solved as

$$\omega = \pm 2\ \sin^{-1}\left[\frac{1}{2}\left(3 - \frac{x_0 - x_{-3}}{x_{-1} - x_{-2}}\right)^{1/2}\right] \qquad \text{(A3.9)}$$

The circular frequency can be chosen to be positive or negative. Either way, the phase angle, ϕ, can be calculated later to agree with the price data. For the sake of convenience, the positive sign is taken in Eqn (A3.9).

Subtracting Eqn (A3.3) from Eqn (A3.2) and after some trigonometric manipulation, we will get

$$2A\cos\phi\,\cos(\omega/2)\sin(\omega/2) + 2A\sin\phi\,\sin(\omega/2) = x_0 - x_{-1} \qquad \text{(A3.10)}$$

Eqn (A3.7) will yield

$$2A\cos\phi\,\cos(3\omega/2)\sin(\omega/2) + 2A\sin\phi\,\sin(3\omega/2)\sin(\omega/2) = x_{-1} - x_{-2}$$

$$\text{(A3.11)}$$

$2A\cos\phi$ and $2A\sin\phi$ can be considered to be two unknowns in Eqn (A3.10) and (A3.11). Solving the two unknowns, we get

$$2A\cos\phi = [(x_0 - x_{-1}) \sin(\omega/2) \sin(3\omega/2) - (x_{-1} - x_{-2}) \sin^2(\omega/2)]/D_0$$

$$\equiv C \qquad (A3.12)$$

$$2A\sin\phi = [(x_{-1} - x_{-2})\sin(\omega/2)\cos(\omega/2) - (x_0 - x_{-1})\sin(\omega/2)\cos(3\omega/2)]/D_0$$

$$\equiv S \qquad (A3.13)$$

where D_0 is given by

$$D_0 = \sin^2(\omega/2)\cos(\omega/2)\sin(3\omega/2) - \sin^3(\omega/2)\cos(\omega/2) \qquad (A3.14)$$

The amplitude A can be calculated from Eqn(A3.13) and (A3.14) as

$$A = [\ (C^2 + S^2)/4\ ]^{1/2} \qquad (A3.15)$$

The phase angle, ϕ, can be calculated from Eqn(A3.13):

$$\phi = \sin^{-1}[\ S/(2A)\ ] \qquad (A3.16)$$

Dividing Eqn (A3.13) by Eqn (A3.12), ϕ can also be calculated as

$$\phi = \tan^{-1}\left[\frac{(x_{-1} - x_{-2})\cos(\omega/2) - (x_0 - x_{-1})\cos(3\omega/2)}{(x_0 - x_{-1})\sin(3\omega/2) - (x_{-1} - x_{-2})\sin(\omega/2)}\right]$$

$$\equiv \tan^{-1}\left[\frac{Y}{X}\right] \qquad (A3.17)$$

where Y is defined as the numerator of the argument of $\tan^{-1}$

and X is defined as the denominator of the argument of $\tan^{-1}$.

It would be better to use Eqn (A3.16) to calculate ϕ, as the amplitude A is seldom equal to 0, and we would have avoided division by 0. If A is by chance equal to 0, it can easily be written in the computer program that the price would equal to the constant level, and ω and ϕ would equal to 0.

However, as quite a number of computer would yield ϕ only between $-\pi/2$ to $\pi/2$, Y and X need to be calculated to determine ϕ between $-\pi$ to π. If $X \geq 0$, the computer program would yield the correct ϕ. If $X < 0$, and $Y \geq 0$, ϕ is calculated by the computer to lie between 0 and $\pi/2$ $(0 \leq \phi \leq \pi/2)$. Our computer program would then set ϕ equal to $\pi - \phi$. If $X < 0$, and $Y < 0$, ϕ is calculated by the computer to lie between $-\pi/2$ and 0 $(-\pi/2 \leq \phi < 0)$. Our computer program would then set ϕ equal to $-(\pi + \phi)$. Thus ϕ would be correctly assigned between $-\pi$ to π.

The constant level, D, can be calculated from Eqn (A3.2):

$$D = x_0 - A\sin\phi = x_0 - S/2 \tag{A3.18}$$

A3.2 Derivation of Frequency (5 points)

When $x_{-1} - x_{-2} = 0$, Eqn (A3.9) cannot be used to calculate ω as it would involve division by 0. One more data point (at $t = -4$) would be required to solve for ω. From Eqn (A3.1)

$$t = -4 \quad x_{-4} = A\sin(-4\omega + \phi) + D \tag{A3.19}$$

Subtracting Eqn (A3.5) from Eqn (A3.3), we get

$$x_{-1} - x_{-3} = A\,[2\cos(-2\omega + \phi)\sin\omega] \tag{A3.20}$$

Subtracting Eqn (A3.19) from Eqn(A3.2), we get

$$x_0 - x_{-4} = A\,[2\cos(-2\omega + \phi)\sin(2\omega)] \tag{A3.21}$$

Dividing Eqn (A3.20) by Eqn (A3.21) will yield

$$(x_0 - x_{-4})/(x_{-1} - x_{-3}) = 2\cos\omega \tag{A3.22}$$

Therefore, ω can be solved as

$$\omega = \cos^{-1}\left[\frac{1}{2}\left(\frac{x_0 - x_{-4}}{x_{-1} - x_{-3}}\right)\right] \tag{A3.23}$$

A3.3 Error Calculation of Frequency (4 points)

Given a function $y = f(x_1, x_2, \ldots\ldots\ldots x_n)$, the error of y, δy, can be calculated as [Bevington 1969]

$$\delta y = \left[\left(\frac{\partial f}{\partial x_1}\right)^2 (\delta x_1)^2 + \left(\frac{\partial f}{\partial x_2}\right)^2 (\delta x_2)^2 + \ldots\ldots\ldots \left(\frac{\partial f}{\partial x_n}\right)^2 (\delta x_n)^2 \right]$$

(A3.24)

where $\partial f/\partial x_i$ are partial derivative with respect to x_i , $i = 1, 2, \ldots..n$.
δx_i are errors of x_i , $i = 1, 2, \ldots..n$.

Calculating the error of ω, $\delta\omega$, in Eqn (A3.9) using Eqn (A3.24), we get

$$\delta\omega = \frac{1}{2}\left[\frac{1}{\left[1 - \frac{1}{4}\left(3 - \frac{x_0 - x_{-3}}{x_{-1} - x_{-2}}\right)\right]\left(3 - \frac{x_0 - x_{-3}}{x_{-1} - x_{-2}}\right)^2} q \right]^{1/2} \quad \text{(A3.25)}$$

where

$$q = \left(\frac{1}{x_{-1} - x_{-2}}\right)^2 \left[(\delta x_0)^2 + (\delta x_{-3})^2\right] + \left(\frac{1}{x_{-1} - x_{-2}}\right)^4 \left[(\delta x_{-1})^2 + (\delta x_{-2})^2\right]$$

(A3.26)

A3.4 Error Calculation of Frequency (5 points)

Calculating the error of ω, δω, in Eqn (A3.23) using Eqn (A3.24), we get

$$\delta\omega = \frac{1}{2}\left[\frac{1}{1-\frac{1}{4}\left(\frac{x_0 - x_{-4}}{x_{-1} - x_{-3}}\right)^2} r\right]^{1/2} \quad \text{(A3.27)}$$

where

$$r = \left(\frac{1}{x_{-1} - x_{-3}}\right)^2 \left[(\delta x_0)^2 + (\delta x_{-4})^2\right] + \left(\frac{1}{x_{-1} - x_{-3}}\right)^4 \left[(\delta x_{-1})^2 + (\delta x_{-3})^2\right]$$

(A3.28)

A3.5 Computer Program for Calculating Frequency

In the EasyLanguage code of Omega Research's TradeStaion2000i, the program for calculating the omega function (ω, which is the circular frequency) and the error of the omega function (δω) using 4 and 5 data points can be written as follows:

```
{omega(Function)}
{S = smoothneses}
Inputs: S(Numeric);
Variables:
omega4(0),omega5(0),den(0),den1(0),sqrt(0),arg(0),sqrtarg(0),dx0(0.01),
dxm1(0.01),dxm2(0.01),dxm3(0.01),dxm4(0.01),f1(0),f2(0),domega4(10
),domega5(10);
{omega4 and omega5 are the circular frequencies calculated using 4 and
5 data points respectively}
{dx0, dxm1, dxm2, dxm3, dxm4 are arbitrary errors assigned to financial
data}
```

{domega4 and domega5, errors of omega4 and omega5 respectively, are assigned arbitrary large numbers in the beginning}

den=(AMAFUNC2(c[1],S)-AMAFUNC2(c[2],S));{Start calculating omega4}

If den=0 Then omega4 = -1; {give omega4 an arbitrary number which will be plotted out of range}

sqrtarg=3-(AMAFUNC2(c,S)-AMAFUNC2(c[3],S))/den;

If sqrtarg < 0 Then omega4 = -2 {give omega4 an arbitray number which will be plotted out of range}

Else Begin

sqrt=Squareroot(sqrtarg);

arg = sqrt/2;

If arg >= 1 Then omega4 = -3 {give omega4 an arbitray number which will be plotted out of range}

Else

omega4 = 2*ArcTangent(arg/Squareroot(1-Square(arg)))*3.14159/180;{omega4 in radians}

omega = omega4;{Assign omega4 to omega first}

If (den=0 OR sqrtarg=0 OR 1-0.25*sqrtarg=0) Then domega4 = 20{Error of omega4 would be large. Arbitrary large error is assigned}

Else Begin

f1=1/(1-0.25*sqrtarg)*1/sqrtarg;{Calculate error of omega4}

f2=Square(1/den)*(dx0*dx0+dxm3*dxm3)+ Power((1/den),4)*(dxm1*dxm1+dxm2*dxm2);

If f1 < 0 Then domega4 = 20 {f1< 0 will yield imaginary number in Squareroot below}

Else

domega4=0.5*Squareroot(f1*f2);

End;

End;

den1=AMAFUNC2(c[1],S)-AMAFUNC2(c[3],S);{Start calculating omega5}

If den1=0 Then omega5 = -4 {give omega5 an arbitray number which will be plotted out of range}

Else Begin

arg= 0.5*(AMAFUNC2(c,S)-AMAFUNC2(c[4],S))/den1;

If AbsValue(arg) > 1 Then omega5 = -5 {give omega5 an arbitray number which will be plotted out of range}

```
Else
        omega5=                    (90-ArcTangent(arg/Squareroot(1-
Square(arg))))*3.14159/180;{omega in radians}
            If (den1=0 OR 1-Square(arg)=0) Then domega5 = 20{Error
of omega5 would be large. Arbitrary large error is assigned}
    Else Begin
        f1=1/(1-Square(arg));{Calculate error of omega5}
        f2=                 Square(1/den1)*(dx0*dx0+dxm4*dxm4)+
Power((1/den1),4)*(dxm1*dxm1+dxm3*dxm3);
        If f1 < 0 Then domega5 = 20 {f1< 0 will yield imaginary number
in Squareroot below}
        Else
        domega5=0.5*Squareroot(f1*f2);
          End;
        End;
        If domega4 >= 20 AND domega5 >= 20 Then omega = -7
{Stringent condition: If domega4 >= 20 OR domega5 >= 20 Then omega
= -7}
        Else Begin
                If domega5 < domega4 Then omega = omega5;{If error
of omega5 is less than that of omeg4, assign omega5 to omega}
                End
```

c represents the closing price of the current data point (bar). c[1] represents the closing price of one bar ago. c[2] represents the closing price of two bars ago. c[3] represents the closing price of three bar ago, and c[4] represents the closing price of four bars ago. AMAFUNC2 is the adaptive moving average function written by Jurik Research. The first input parameter of AMAFUNC2 signifies the closing price series to be smoothed, while the second input parameter indicates the smoothness factor, S, which is arranged to be a variable decided by the user. For example, S = 32 and 3 in Fig 6.5 and 6.6 respectively. The larger the smoothness factor, the more smoothed the smoothed data will be. AMAFUNC2 can be substituted by other smoothing function. For example, it can be substituted by XAVERAGE, which is a build-in exponential moving average function written by TradeStation2000i.

The omega function program calculates the value of the circular frequency, ω, and its error using 4 points. It also calculates the value of the circular frequency, ω, and its error using 5 points. The two errors are compared, the ω which has the lesser error will be plotted. For those 4 *and* 5 data points that cannot fit into the sine wave model, the error of ω is arbitrarily set to be 20, which is a large number compared to the usual calculated error. The circular frequency, ω is then given an arbitrary negative number, so that it would not be plotted in the positive scale determined by the user (Fig 6.5 and 6.6). The indicator program for plotting omega is listed below:

```
{omegaplot(Indicator)}
Inputs: S(3);
Plot1(omega(S),"Plot1");
```

In the above program, the smoothness parameter, S, is set to be 3 by default. This input parameter can be changed to other positive integer. This indicator program calls the omega function, and plots it as dots (or a line if the user so chooses) in the figure (see bottom plots of Fig. 6.5 and 6.6).

A3.6 Computer Programs for Calculating Wave Velocity and Wave Acceleration

To plot the wave velocity and acceleration, the amplitude of the sine wave, A, as well as its phase when $t = 0$, ϕ, need to be calculated. In the EasyLanguage code of Omega Research's TradeStaion2000i, the program for calculating the amplitude function, called Ampprice, is listed below:

```
{Ampprice(function)}
Inputs: S(Numeric);
If omega(S) < 0  Then Ampprice=-1 {Ampprice is assigned a negative
number}
Else
        Ampprice=Squareroot((Square(TAcosphi(S))+Square(TAsinphi(
S)))/4)
```

The amplitude is calculated from Eqn (A3.15). The program calls two function programs, TAcosphi and TAsinphi, which are listed as follows:

```
{TAcosphi (function) }
Inputs: S(Numeric);
Variables: sin(0),cos(0),sin3(0),cos3(0),D0(0);
        sin=Sine(omega(S)*(180/3.14159)/2);
   cos=Cosine(omega(S)*(180/3.14159)/2);
        sin3=Sine(3*omega(S)*(180/3.14159)/2);
        cos3=Cosine(3*omega(S)*(180/3.14159)/2);
        D0=sin*sin*cos*sin3-sin*sin*sin*cos3;
   If D0=0 Then D0=0.00000001;{To avoid dividing by zero}
        TAcosphi=(1/D0)*((AMAFUNC2(c,S)-
AMAFUNC2(c[1],S))*sin*sin3-(AMAFUNC2(c[1],S)-
AMAFUNC2(c[2],S))*sin*sin);
```

```
{TAsinphi (function) }
Inputs: S(Numeric);
Variables: sin(0),cos(0),sin3(0),cos3(0),D0(0);
        sin=Sine(omega(S)*(180/3.14159)/2);
   cos=Cosine(omega(S)*(180/3.14159)/2);
        sin3=Sine(3*omega(S)*(180/3.14159)/2);
        cos3=Cosine(3*omega(S)*(180/3.14159)/2);
        D0=sin*sin*cos*sin3-sin*sin*sin*cos3;
   If D0=0 Then D0=0.00000001;{To avoid dividing by zero}
        TAsinphi=(1/D0)*((AMAFUNC2(c[1],S)-
AMAFUNC2(c[2],S))*sin*cos-(AMAFUNC2(c,S)-
AMAFUNC2(c[1],S))*sin*cos3);
```

The program for calculating the phase when $t = 0$, ϕ, the phi function is listed below:

```
{phi (function) }
Inputs: S(Numeric);
Variables: sin(0),cos(0),sin3(0),cos3(0),Y(0),X(0),phidegree(0);
If omega(S) < 0 Then phi=-5 { phi assigned a negative number so that it
can be plotted out of range}
Else Begin      sin=Sine(omega(S)*(180/3.14159)/2);
```

```
   cos=Cosine(omega(S)*(180/3.14159)/2);
         sin3=Sine(3*omega(S)*(180/3.14159)/2);
         cos3=Cosine(3*omega(S)*(180/3.14159)/2);
         Y=(AMAFUNC2(c[1],S)-AMAFUNC2(c[2],S))*cos-
(AMAFUNC2(c,S)-AMAFUNC2(c[1],S))*cos3;
         X=(AMAFUNC2(c,S)-AMAFUNC2(c[1],S))*sin3-
(AMAFUNC2(c[1],S)-AMAFUNC2(c[2],S))*sin;
   If X=0 Then X=0.000000001;{to avoid dividing by zero}
   phidegree=AbsValue(ArcTangent(Y/X));
   If X < 0 AND Y > 0 Then phidegree = 180-phidegree;
   If X < 0 AND Y < 0 Then phidegree = 180+phidegree;
   If X > 0 AND Y < 0 Then phidegree = 360-phidegree;
         phi=phidegree;
         End
```

As described in this Appendix, to calculate ϕ, it would be more accurate to use arcsine instead of arctangent. However, since TradeStation2000i does not have a build-in arcsine function, arctangent is used instead. In any case, the numerator and denominator of the argument in the arctangent function needs to be used to calculate which quadrant the phase angle is in.

Knowing ω, A and ϕ, wave velocity can be calculated and plotted. The program for calculating and plotting the wave velocity indicator, velfit, is listed below:

```
{velfit(Indicator)}
Inputs: S(3);
Variables: velfit(0);
If omega(S) < 0  Then velfit=200 {velfit=200 will be plotted out of
scale}
Else Begin
         velfit=Ampprice(S)*omega(S)*Cosine(phi(S));
         End;
Plot1(velfit,"Plot1");
```

The smoothness factor, S, is set to 3 by default, but it can be changed as it is designed to be an input parameter to the program.

Again, knowing ω, A and ϕ, wave acceleration can be calculated and plotted. The program for calculating and plotting the wave acceleration indicator, accelfit, is listed below:

```
{accelfit(Indicator)}
Inputs: S(3);
Variables: accelfit(0);
If omega(S) < 0  Then accelfit=100 {accelfit=100 will be plotted out of
scale}
Else Begin
        accelfit=-Ampprice(S)*Square(omega(S))*Sine(phi(S));
        End;
Plot1(accelfit,"Plot1");
```

The constant level, D, can be calculated if necessary and then plotted. The function program for calculating the constant level is listed below:

```
{Dclevel(function) }
Inputs: S(Numeric);
If omega(S) < 0  Then Dclevel= -1
Else
                Dclevel=AMAFUNC2(c,S)-TAsinphi(S)/2
```

The function program for plotting the constant level is listed below:

```
{Dclevelplot(Indicator)}
Inputs: S(3);
Plot1(Dclevel(S),"Plot1");
```

Appendix 4

Higher Order Polynomial High Pass Filters

A4.1 Derivation of Quartic Indicators

A4.1.1 *Quartic Velocity Indicator*

A quartic function is a fourth order polynomial of the form:

$$x(t) = at^4 + bt^3 + ct^2 + dt + e \qquad \text{(A4.1)}$$

where t is a continuous variable,

a, b, c, d and e are constant coefficients.

For discrete time signals, Eq (A4.1) can be written as

$$x(n) = an^4 + bn^3 + cn^2 + dn + e \qquad \text{(A4.2)}$$

where n is an integer.

We are interested to find the derivative of the quartic function at n = 0, which is the most recent data point.

From Eq (A4.2), we write

$$x_0 = x(0) = e \qquad \text{(A4.3a)}$$

$$x_{-1} = x(-1) = a - b + c - d + e \qquad \text{(A4.3b)}$$

$$x_{-2} = x(-2) = 16a - 8b + 4c - 2d + e \qquad \text{(A4.3c)}$$

$x_{-3} = x(-3) = 81a - 27b + 9c - 3d + e$ (A4.3d)

$x_{-4} = x(-4) = 256a - 64b + 16c - 4d + e$ (A4.3e)

Solving Eq (A4.3a-e) for a, b, c, d and e by using determinants, we get

$a = (1/24)x_0 - (1/6)x_{-1} + (1/4)x_{-2} - (1/6)x_{-3} + (1/24)x_{-4}$ (A4.4a)

$b = (5/12)x_0 - (3/2)x_{-1} + 2x_{-2} - (7/6)x_{-3} + (1/4)x_{-4}$ (A4.4b)

$c = (35/24)x_0 - (13/3)x_{-1} + (19/4)x_{-2} - (7/3)x_{-3} + (11/24)x_{-4}$ (A4.4c)

$d = (25/12)x_0 - 4x_{-1} + 3x_{-2} - (4/3)x_{-3} + (1/4)x_{-4}$ (A4.4d)

$e = x_0$ (A4.4e)

Taking the derivative of Eq (A4.1) and then substitute t with n, we arrive at

$$\frac{dx}{dn} = 4an^3 + 3bn^2 + 2cn + d \quad \text{(A4.5)}$$

At n = 0,

$$\left.\frac{dx}{dn}\right|_{n=0} = d = \frac{25}{12}x_0 - 4x_{-1} + 3x_{-2} - \frac{4}{3}x_{-3} + \frac{1}{4}x_{-4} \quad \text{(A4.6)}$$

We will define the unit sample response h of the quartic velocity indicator as

$h = (h(0), h(1), h(2), h(3), h(4)) = (25/12, -4, 3, -4/3, 1/4)$ (A4.7)

Thus the output response is given by the convolution sum

$$y(n) = \frac{25}{12}x(n) - 4x(n-1) + 3x(n-2) - \frac{4}{3}x(n-3) + \frac{1}{4}x(n-4) \quad \text{(A4.8)}$$

The frequency response or the Discrete Time Fourier Transform (DTFT) of h is given by

$H(\omega) =$

$25/12 - 4\exp(-i\omega) + 3\exp(-2i\omega) - (4/3)\exp(-3i\omega) + (1/4)\exp(-4i\omega)$

(A4.9)

A4.1.2 *Quartic Acceleration Indicator*

Taking the second derivative of Eq (A4.1) and then substitute t with n, we arrive at

$$\frac{d^2x}{dn^2} = 12an^2 + 6bn + 2c \qquad \text{(A4.10)}$$

At n = 0,

$$\left.\frac{d^2x}{dn^2}\right|_{n=0} = 2c = \frac{35}{12}x_0 - \frac{26}{3}x_{-1} + \frac{19}{2}x_{-2} - \frac{14}{3}x_{-3} + \frac{11}{12}x_{-4} \qquad \text{(A4.11)}$$

We will define the unit sample response h of the quartic acceleration indicator as

h = (h(0), h(1), h(2), h(3), h(4))

= (35/12, -26/3, 19/2, -14/3, 11/12) (A4.12)

Thus the output response is given by the convolution sum

$$y(n) = \frac{35}{12}x(n) - \frac{26}{3}x(n-1) + \frac{19}{2}x(n-2) - \frac{14}{3}x(n-3) + \frac{11}{12}x(n-4)$$

(A4.13)

The frequency response or the Discrete Time Fourier Transform (DTFT) of h is given by

$H(\omega) =$

$$35/12 - (26/3)\exp(-i\omega) + (19/2)\exp(-2i\omega) - (14/3)\exp(-3i\omega) + (11/12)\exp(-4i\omega) \quad \text{(A4.14)}$$

A4.2 Derivation of Quintic Indicators

A4.2.1 *Quintic Velocity Indicator*

A quintic function is a fifth order polynomial of the form:

$$x(t) = at^5 + bt^4 + ct^3 + dt^2 + et + f \quad \text{(A4.15)}$$

where t is a continuous variable,

a, b, c, d, e and f are constant coefficients.

For discrete time signals, Eq (A4.15) can be written as

$$x(n) = an^5 + bn^4 + cn^3 + dn^2 + en + f \quad \text{(A4.16)}$$

where n is an integer.

We are interested to find the derivative of the quintic function at n = 0, which is the most recent data point.

From Eq (A4.16), we write

$$x_0 = x(0) = f \quad \text{(A4.17a)}$$

$$x_{-1} = x(-1) = -a + b - c + d - e + f \quad \text{(A4.17b)}$$

$$x_{-2} = x(-2) = -32a + 16b - 8c + 4d - 2e + f \quad \text{(A4.17c)}$$

$$x_{-3} = x(-3) = -243a + 81b - 27c + 9d - 3e + f \quad \text{(A4.17d)}$$

$$x_{-4} = x(-4) = -1024a + 256b - 64c + 16d - 4e + f \quad \text{(A4.17e)}$$

$$x_{-5} = x(-5) = -3125a + 625b - 125c + 25d - 5e + f \quad \text{(A4.17f)}$$

Solving Eq (A4.17a-f) for a, b, c, d, e and f by using determinants, we get

$$a = (1/120)x_0 - (1/24)x_{-1} + (1/12)x_{-2} - (1/12)x_{-3} + (1/24)x_{-4} - (1/120)x_{-5} \quad \text{(A4.18a)}$$

$$b = (1/8)x_0 - (7/12)x_{-1} + (13/12)x_{-2} - x_{-3} + (11/24)x_{-4} - (1/12)x_{-5} \quad \text{(A4.18b)}$$

$$c = (17/24)x_0 - (71/24)x_{-1} + (59/12)x_{-2} - (49/12)x_{-3} + (41/24)x_{-4} - (1/12)x_{-5} \quad \text{(A4.18c)}$$

$$d = (15/8)x_0 - (77/12)x_{-1} + (107/12)x_{-2} - (13/2)x_{-3} + (61/24)x_{-4} - (5/12)x_{-5} \quad \text{(A4.18d)}$$

$$e = (137/60)x_0 - 5x_{-1} + 5x_{-2} - (10/3)x_{-3} + (5/4)x_{-4} - (1/5)x_{-5} \quad \text{(A4.18e)}$$

$$f = x_0 \quad \text{(A4.18f)}$$

Taking the derivative of Eq (A4.15) and then substitute t with n, we arrive at

$$\frac{dx}{dn} = 5an^4 + 4bn^3 + 3cn^2 + 2dn + e \quad \text{(A4.19)}$$

At $n = 0$,

$$\left.\frac{dx}{dn}\right|_{n=0} = e = \frac{137}{60}x_0 - 5x_{-1} + 5x_{-2} - \frac{10}{3}x_{-3} + \frac{5}{4}x_{-4} - \frac{1}{5}x_{-5} \qquad \text{(A4.20)}$$

We will define the unit sample response h of the quintic velocity indicator as

$h = (h(0), h(1), h(2), h(3), h(4), h(5))$

$= (137/60,\ -5,\ 5,\ -10/3,\ 5/4,\ -1/5)$ (A4.21)

Thus the output response is given by the convolution sum

$$y(n) = \frac{137}{60}x(n) - 5x(n-1) + 5x(n-2) - \frac{10}{3}x(n-3) + \frac{5}{4}x(n-4) - \frac{1}{5}x(n-5) \qquad \text{(A4.22)}$$

The frequency response or the Discrete Time Fourier Transform (DTFT) of h is given by

$$H(\omega) = 137/60 - 5\exp(-i\omega) + 5\exp(-2i\omega) - (10/3)\exp(-3i\omega) + (5/4)\exp(-4i\omega) - (1/5)\exp(-5i\omega) \qquad \text{(A4.23)}$$

A4.2.2 *Quintic Acceleration Indicator*

Taking the second derivative of Eq (A4.15) and then substitute t with n, we arrive at

$$\frac{d^2x}{dn^2} = 20an^3 + 12bn^2 + 6cn + 2d \qquad \text{(A4.24)}$$

At n = 0,

$$\left.\frac{d^2x}{dn^2}\right|_{n=0} = 2d = \frac{15}{4}x_0 - \frac{77}{6}x_{-1} + \frac{107}{6}x_{-2} - 13x_{-3} + \frac{61}{12}x_{-4} - \frac{5}{6}x_{-5} \tag{A4.25}$$

We will define the unit sample response h of the quintic acceleration indicator as

$$h = (h(0), h(1), h(2), h(3), h(4), h(5))$$
$$= (15/4,\ -77/6,\ 107/6,\ -13,\ 61/12,\ -5/6) \tag{A4.26}$$

Thus the output response is given by the convolution sum

$$y(n) = \frac{15}{4}x(n) - \frac{77}{6}x(n-1) + \frac{107}{6}x(n-2) - 13x(n-3) + \frac{61}{12}x(n-4) - \frac{5}{6}x(n-5) \tag{A4.27}$$

The frequency response or the Discrete Time Fourier Transform (DTFT) of h is given by

$$H(\omega) = 15/4 - (77/6)\exp(-i\omega) - (107/6)\exp(-2i\omega) - 13\exp(-3i\omega) + (61/12)\exp(-4i\omega) - (5/6)\exp(-5i\omega) \tag{A4.28}$$

A4.3 Derivation of Sextic Indicators

A4.3.1 *Sextic Velocity Indicator*

A sextic function is a sixth order polynomial of the form:

$$x(t) = at^6 + bt^5 + ct^4 + dt^3 + et^2 + ft + g \tag{A4.29}$$

where t is a continuous variable,

a, b, c, d, e, f and g are constant coefficients.

For discrete time signals, Eq (A4.15) can be written as

$$x(n) = an^6 + bn^5 + cn^4 + dn^3 + en^2 + fn + g \quad \text{(A4.30)}$$

where n is an integer.

We are interested to find the derivative of the sextic function at n = 0, which is the most recent data point.

From Eq (A4.30), we write

$$x_0 = x(0) = g \quad \text{(A4.31a)}$$

$$x_{-1} = x(-1) = a - b + c - d + e - f + g \quad \text{(A4.31b)}$$

$$x_{-2} = x(-2) = 64a - 32b + 16c - 8d + 4e - 2f + g \quad \text{(A4.31c)}$$

$$x_{-3} = x(-3) = 729a - 243b + 81c - 27d + 9e - 3f + g \quad \text{(A4.31d)}$$

$$x_{-4} = x(-4) = 4096a - 1024b + 256c - 64d + 16e - 4f + g \quad \text{(A4.31e)}$$

$$x_{-5} = x(-5) = 15625a - 3125b + 625c - 125d + 25e - 5f + g \quad \text{(A4.31f)}$$

$$x_{-6} = x(-6) = 46656a - 7776b + 1296c - 216d + 36e - 6f + g \quad \text{(A4.31g)}$$

Solving Eq (A4.31a-g) for a, b, c, d, e, f and g by using determinants, we get

$$a = (34560\, x_0 - 207360\, x_{-1} + 518400\, x_{-2} - 691200\, x_{-3} + 518400\, x_{-4} - 207360\, x_{-5} + 34560\, x_{-6}) / 24883200 \quad \text{(A4.32a)}$$

$$b = (725760\, x_0 - 4147200\, x_{-1} + 9849600\, x_{-2} - 12441600\, x_{-3} + 8812800\, x_{-4} - 3317760\, x_{-5} + 518400\, x_{-6}) / 24883200 \quad \text{(A4.32b)}$$

$$c = (6048000\, x_0 - 32140800\, x_{-1} + 71020800\, x_{-2} - 83635200\, x_{-3} + 55468800\, x_{-4} - 19699200\, x_{-5} + 2937600\, x_{-6}) / 24883200 \quad \text{(A4.32c)}$$

$$d = (25401600\, x_0 - 120268800\, x_{-1} + 238982400\, x_{-2} - 257126400 x_{-3} + 159148800\, x_{-4} - 53913600\, x_{-5} + 7776000\, x_{-6}) / 24883200 \quad \text{(A4.32d)}$$

$$e = (56125440\, x_0 - 216483840\, x_{-1} + 363916800\, x_{-2} - 351129600 x_{-3} + 205286400\, x_{-4} - 67184640\, x_{-5} + 9469440\, x_{-6}) / 24883200 \quad \text{(A4.32e)}$$

$$f = (49/20)\, x_0 - 6\, x_{-1} + (15/2)\, x_{-2} - (20/3)\, x_{-3} + (15/4)\, x_{-4} - (6/5)\, x_{-5} + (1/6)\, x_{-6} \quad \text{(A4.32f)}$$

$$g = x_0 \quad \text{(A4.32g)}$$

Taking the derivative of Eq (A4.29) and then substitute t with n, we arrive at

$$\frac{dx}{dn} = 6an^5 + 5bn^4 + 4cn^3 + 3dn^2 + 2en + f \quad \text{(A4.33)}$$

At $n = 0$,

$$\left.\frac{dx}{dn}\right|_{n=0} = f = \frac{49}{20}x_0 - 6x_{-1} + \frac{15}{2}x_{-2} - \frac{20}{3}x_{-3} + \frac{15}{4}x_{-4} - \frac{6}{5}x_{-5} + \frac{1}{6}x_{-6} \tag{A4.34}$$

We will define the unit sample response h of the sextic velocity indicator as

$h = (h(0), h(1), h(2), h(3), h(4), h(5), h(6))$

$= (49/20,\ -6,\ 15/2,\ -20/3,\ 15/4,\ -6/5,\ 1/6)$ (A4.35)

Thus the output response is given by the convolution sum

$$y(n) = \frac{49}{20}x(n) - 6x(n-1) + \frac{15}{2}x(n-2) - \frac{20}{3}x(n-3) + \frac{15}{4}x(n-4)$$

$$- \frac{6}{5}x(n-5) + \frac{1}{6}x(n-6) \tag{A4.36}$$

The frequency response or the Discrete Time Fourier Transform (DTFT) of h is given by

$$H(\omega) = 49/20 - 6\exp(-i\omega) + (15/2)\exp(-2i\omega) - (20/3)\exp(-3i\omega)$$

$$+(15/4)\exp(-4i\omega) - (6/5)\exp(-5i\omega) + (1/6)\exp(-6i\omega) \tag{A4.37}$$

A4.3.2 *Sextic Acceleration Indicator*

Taking the second derivative of Eq (A4.29) and then substitute t with n, we arrive at

$$\frac{d^2x}{dn^2} = 30an^4 + 20bn^3 + 12cn^2 + 6dn + 2e \tag{A4.38}$$

At n = 0,

$$\left.\frac{d^2x}{dn^2}\right|_{n=0} = 2e = 4.5112x_0 - 17.4x_{-1} + 29.25x_{-2} - 28.2222x_{-3}$$

$$+16.5x_{-4} - 5.4x_{-5} + 0.7612x_{-6} \qquad \text{(A4.39)}$$

We will define the unit sample response h of the quintic acceleration indicator as

$$h = (h(0), h(1), h(2), h(3), h(4), h(5), h(6))$$

$$= (4.5112,\ -17.4,\ 29.25,\ -28.2222,\ 16.5,\ -5.4,\ 0.7612) \qquad \text{(A4.40)}$$

Thus the output response is given by the convolution sum

$$y(n) = 4.5112x(n) - 17.4x(n-1) + 29.25x(n-2) - 28.2222x(n-3)$$

$$+16.5x(n-4) - 5.4x(n-5) + 0.7612x(n-6) \qquad \text{(A4.41)}$$

The frequency response or the Discrete Time Fourier Transform (DTFT) of h is given by

$$H(\omega) = 4.5112 - 17.4\exp(-i\omega) - 29.25\exp(-2i\omega) - 28.2222\exp(-3i\omega)$$

$$+ 16.5\exp(-4i\omega) - 5.4\exp(-5i\omega) + 0.7612\exp(-6i\omega) \qquad \text{(A4.42)}$$

Appendix 5

MATLAB Programs for Money Management

The following two programs are written to calculate probability, expected value, total probability, total expected value and total expected value/average time of a trade, with a trailing stop-loss or fixed stop-loss. They are not necessarily written in the most efficient manner.

The parameters used in the Levy distribution in the programs correspond to the S & P 500 index. Those parameters need to be changed for calculation of other markets.

A5.1 Trailing Stop-Loss Program

```
%Trailingstoploss,   t = 10
%
clear
t=10;
for I=1:t+1
%day is only an example of one time unit
   plossvector(I)=0;% probability of that day's
stopping out by initial stoploss or trailing stoploss
   Exlossvector(I)=0;% expected value of being
stopped out
   pvaluevector(I)=0;% probability of last day' cash
out
   Totalp(I)=0;% Total probability of previous days'
stops and last day cash out
   Totalex(I)=0;
   Expectedt(I)=0;
end
deltat=1;
```

```
interval = .05;% for integral of Eqn (1) in Mantagna
and Stanley Nature V376 P48
endpoint = 500;% for integral of Eqn (1) in Mantagna
and Stanley Nature V376 P48
q = 0: interval: endpoint;
zinterval = 0.0125;  % can change zinterval
zendpoint = 0.075;% can change zendpoint
z = 0: zinterval: zendpoint;
Nptlevy = round(zendpoint/zinterval) +1 ;
for J=1:Nptlevy
   for I=1:endpoint/interval +1
   y(I)= (1/pi)*exp(-
0.00375*deltat*q(I)^1.4)*cos(q(I)*z(J));
   end
   Lalpha(J)=trapz(q,y); %calculate integral using
trapezoidal method,
end
figure(1)
plot(z,Lalpha,'k*-')
xlabel('Z,   Delta t = 1')
ylabel('Levy distribution')  %   Eqn (1) in Mantagna
and Stanley paper, Nature V376 P48
% To assign numbers to the symmetrical Lalphanegpov
function
for J=1: 2*Nptlevy -1
   znegpos(J)=0;
   Lalphanegpos(J)=0;
end
   for J = 1:Nptlevy - 1
   znegpos(J) = -z(Nptlevy+1 - J);
   Lalphanegpos(J) = Lalpha(Nptlevy+1 - J);
end
for J= 1: Nptlevy
   znegpos(J + Nptlevy -1) = z(J);
   Lalphanegpos(J + Nptlevy -1) = Lalpha(J);
end
figure(2)
plot(znegpos,Lalphanegpos,'k*-')
xlabel('Z,   Delta t = 1')
ylabel('L,   Eqn (1) in paper')
Sumlevy=0;% normalize probability distribution
for I=1:2*Nptlevy -1
   Sumlevy = Sumlevy + Lalphanegpos(I);
end
```

```
for I=1:2*Nptlevy -1
   Lalphanegpos(I) = Lalphanegpos(I)/Sumlevy;
end
figure(3)
plot(znegpos,Lalphanegpos,'k*-')
xlabel('Z,   Delta t = 1')
ylabel('Normalized Levy distribution')
title('Normalized probability distribution for S & P
500')
%Nptlevy = 2;% for testing only
%znegpos=[-1 0 1 ];% for testing only
%Lalphanegpos = [ 0.25  0.5  0.25];% for testing only
%Nptlevy = 3;% for testing only
%znegpos=[-2 -1 0 1 2];% for testing only
%Lalphanegpos = [0.1  0.2  0.4  0.2  0.1];% for
testing only
%Nptlevy = 4;% for testing only
%znegpos=[-3 -2 -1 0 1 2 3];% for testing only
%Lalphanegpos = [0.1  0.125  .175  .2  .175  .125
.1];% for testing only
%Lalphanegpos = [0.05  0.1  0.2  0.3  0.2  0.1
0.05];% for testing only
S= -2;%                   change stoploss   S = 0, -1,
-2, ...  ***********
% Setting S = -1 will yield a constant TotalexDt with
respect to t
znegposno = Nptlevy + S;% this is the number in the
vector where stoploss is set
stoploss=znegpos(znegposno);
% Buy at t = 0, to find the expectation value at
furture t
Exvaluevector(1) = 0; % at time t=0,  buy --
therefore no gain nor loss
t=1% t = 1 time unit
tvector(t) = t;
totalprobstoploss=0;
for K = 1:znegposno
   totalprobstoploss = totalprobstoploss +
Lalphanegpos(K);
end
% totalprobstoploss is the sum of the probabilities
that the trade will be stopped
Exloss1 = totalprobstoploss * znegpos(znegposno) %
znegpos(znegposno)= stop loss value
```

```
pvalue1=1;% this is the sum of the normalized
Lalphanegpo.  That is why it is equal to 1
Exvalue1=Exloss1;
for K = znegposno+1:2*Nptlevy -1
   Exvalue1 = Exvalue1 + znegpos(K)*Lalphanegpos(K)
end
plossvector(2)=totalprobstoploss;%probloss=
probalility of being stopped at an initial stoploss
or trailing stop loss.  For a trailing stoploss,
there can be profit and not loss.
Exlossvector(t+1) = Exloss1; % expected loss or gain
at t = 1
pvaluevector(t+1) = pvalue1;% this number is not used
Exvaluevector(t+1) = Exvalue1; % expected value at t
= 1
t=2 % t = 2 time units
ploss1=0;
pvalue1=0;
Exloss1=0;
Exvalue1=0;
for K1 = znegposno +1: 2*Nptlevy -1% 2*Nptlevy -1 = m
in Eqn
       T1 = znegpos(K1);
       ploss2=0;
       Exloss2=0;
       pvalue2=0;
       Exvalue2 = 0;

    for K2 = 1:2*Nptlevy -1 % znegposno = -n in Eqn
          T2 = znegpos(K1)+ znegpos(K2);
          Dr=1;
          if T1<= 0, Dr=0; end
          Ds =1;
      if T2 <= stoploss , Ds = 0;end
      Dt=1;
      if T2 <= T1 +stoploss, Dt=0; end% Dt relates to
trailing stoploss
      Exvalue2 = Exvalue2 + Lalphanegpos(K2)*((1-
Dr)*stoploss*(1-Dt)+Dr*(T1+stoploss)*(1-Dt)+(1-
Ds)*stoploss*Dt+Ds*T2*Dt);
      ploss2 = ploss2 + Lalphanegpos(K2)*((1-Dr)*(1-
Dt)+Dr*(1-Dt)+(1-Ds)*Dt);
```

```
      Exloss2 = Exloss2 + Lalphanegpos(K2)*((1-
Dr)*stoploss*(1-Dt)+Dr*(T1+stoploss)*(1-Dt)+(1-
Ds)*stoploss*Dt);
      pvalue2 = 1;% pvalue2 adds up to  1
    end
    ploss1= ploss1+Lalphanegpos(K1)*ploss2
    Exloss1= Exloss1+Lalphanegpos(K1)*Exloss2
    pvalue1 = pvalue1 + Lalphanegpos(K1)*pvalue2
    Exvalue1 = Exvalue1 + Lalphanegpos(K1)*Exvalue2
 end
 plossvector(t+1)=ploss1;% probability of the trade
being stopped out at t=2, and that it is not stopped
at t=1
 pvaluevector(t+1) = pvalue1;%prob of trader cashing
out at t=2, and that the trade is not stopped at t=1
 Exlossvector(t+1) = Exloss1;%Expected loss(or maybe
gain) at t=2 when the trade is stopped out at t=2
 Exvaluevector(t+1) = Exvalue1;%Expected value at
t=2, given that the trade is not stopped at t=1
 t=3
 ploss1=0;
 pvalue1=0;
 Exloss1=0;
 Exvalue1=0;
 for K1 = znegposno +1: 2*Nptlevy -1
       T1 = znegpos(K1);
       ploss2=0;
       pvalue2=0;
       Exloss2=0;
       Exvalue2 = 0;
   for K2 = znegposno +1: 2*Nptlevy -1
       T2 = znegpos(K1)+ znegpos(K2);
       ploss3=0;
       Exloss3=0;
       pvalue3=0;
       Exvalue3 = 0;

   for K3 = 1:2*Nptlevy -1 % znegposno = -n in Eqn
          T3 = znegpos(K1)+ znegpos(K2)+ znegpos(K3);
          Dr=1;
          if T2<= 0, Dr=0; end
          Ds =1;
      if T3 <= stoploss , Ds = 0;end
      Dt=1;
```

```
      if T3 <= T2 +stoploss, Dt=0; end% Dt relates to
trailing stoploss
      Exvalue3 = Exvalue3 + Lalphanegpos(K3)*((1-
Dr)*stoploss*(1-Dt)+Dr*(T2+stoploss)*(1-Dt)+(1-
Ds)*stoploss*Dt+Ds*T3*Dt);
      ploss3 = ploss3 + Lalphanegpos(K3)*((1-Dr)*(1-
Dt)+Dr*(1-Dt)+(1-Ds)*Dt);
      Exloss3 = Exloss3 + Lalphanegpos(K3)*((1-
Dr)*stoploss*(1-Dt)+Dr*(T2+stoploss)*(1-Dt)+(1-
Ds)*stoploss*Dt);
      pvalue3 = 1;
    end

       D =1;
       if T2 <= znegpos(znegposno) , D = 0;end
       ploss2 = ploss2 + D * Lalphanegpos(K2)*ploss3;
       pvalue2 = pvalue2 + D *
Lalphanegpos(K2)*pvalue3;
       Exloss2 = Exloss2 + D *
Lalphanegpos(K2)*Exloss3;
       Exvalue2 = Exvalue2 + D *
Lalphanegpos(K2)*Exvalue3;
   end
 ploss1 = ploss1+ Lalphanegpos(K1)*ploss2
 pvalue1 = pvalue1+ Lalphanegpos(K1)*pvalue2
 Exloss1 = Exloss1+ Lalphanegpos(K1)*Exloss2
 Exvalue1 = Exvalue1+ Lalphanegpos(K1)*Exvalue2
end
plossvector(t+1) = ploss1;
pvaluevector(t+1) = pvalue1;
Exlossvector(t+1) = Exloss1;
Exvaluevector(t+1) = Exvalue1;
t=4
ploss1=0;
pvalue1=0;
Exloss1=0;
Exvalue1=0;
 for K1 = znegposno +1: 2*Nptlevy -1
       T1 = znegpos(K1);
       ploss2=0;
       pvalue2 = 0;
       Exloss2=0;
       Exvalue2 = 0;
      for K2 = znegposno +1: 2*Nptlevy -1
```

```
        T2 = znegpos(K1)+ znegpos(K2);
        ploss3=0;
        pvalue3 = 0;
        Exloss3=0;
        Exvalue3 = 0;
       for K3 = znegposno +1: 2*Nptlevy -1
        T3 = znegpos(K1)+ znegpos(K2)+ znegpos(K3);
        ploss4=0;
        Exloss4 = 0;
        pvalue4=0;
        Exvalue4 = 0;

       for K4 = 1:2*Nptlevy -1 % znegposno = -n in Eqn
           T4 = znegpos(K1)+ znegpos(K2)+ znegpos(K3)+
znegpos(K4);
           Dr=1;
           if T3<= 0, Dr=0; end
           Ds =1;
       if T4 <= stoploss , Ds = 0;end
       Dt=1;
       if T4 <= T3 +stoploss, Dt=0; end% Dt relates to
trailing stoploss
       Exvalue4 = Exvalue4 + Lalphanegpos(K4)*((1-
Dr)*stoploss*(1-Dt)+Dr*(T3+stoploss)*(1-Dt)+(1-
Ds)*stoploss*Dt+Ds*T4*Dt);
       ploss4 = ploss4 + Lalphanegpos(K4)*((1-Dr)*(1-
Dt)+Dr*(1-Dt)+(1-Ds)*Dt);
       Exloss4 = Exloss4 + Lalphanegpos(K4)*((1-
Dr)*stoploss*(1-Dt)+Dr*(T3+stoploss)*(1-Dt)+(1-
Ds)*stoploss*Dt);
       pvalue4 = 1;
     end

      D =1;
        if T3 <= znegpos(znegposno) , D = 0;end
        ploss3 = ploss3 + D * Lalphanegpos(K3)*ploss4;
        pvalue3 = pvalue3 + D *
Lalphanegpos(K3)*pvalue4;
        Exloss3 = Exloss3 + D *
Lalphanegpos(K3)*Exloss4;
        Exvalue3 = Exvalue3 + D *
Lalphanegpos(K3)*Exvalue4;
      end
        D =1;
```

```
       if T2 <= znegpos(znegposno) , D = 0;end
       ploss2 = ploss2 + D * Lalphanegpos(K2)*ploss3;
       pvalue2 = pvalue2 + D *
Lalphanegpos(K2)*pvalue3;
       Exloss2 = Exloss2 + D *
Lalphanegpos(K2)*Exloss3;
       Exvalue2 = Exvalue2 + D *
Lalphanegpos(K2)*Exvalue3;
 end
 ploss1 = ploss1+ Lalphanegpos(K1)*ploss2;
 pvalue1 = pvalue1+ Lalphanegpos(K1)*pvalue2;
 Exloss1 = Exloss1+ Lalphanegpos(K1)*Exloss2;
 Exvalue1 = Exvalue1+ Lalphanegpos(K1)*Exvalue2;
end
plossvector(t+1) = ploss1;
pvaluevector(t+1) = pvalue1;
Exlossvector(t+1) = Exloss1;
Exvaluevector(t+1) = Exvalue1;
t=5
ploss1=0;
pvalue1=0;
Exloss1=0;
Exvalue1=0;
 for K1 = znegposno +1: 2*Nptlevy -1
       T1 = znegpos(K1);
       ploss2=0;
       pvalue2 = 0;
       Exloss2=0;
       Exvalue2 = 0;
      for K2 = znegposno +1: 2*Nptlevy -1
       T2 = znegpos(K1)+ znegpos(K2);
       ploss3=0;
       pvalue3 = 0;
       Exloss3=0;
       Exvalue3 = 0;
      for K3 = znegposno +1: 2*Nptlevy -1
       T3 = znegpos(K1)+ znegpos(K2)+ znegpos(K3);
       ploss4=0;
       pvalue4 = 0;
       Exloss4=0;
       Exvalue4 = 0;
      for K4 = znegposno +1: 2*Nptlevy -1
       T4 = znegpos(K1)+ znegpos(K2)+ znegpos(K3)+
znegpos(K4);
```

```
        ploss5=0;
        Exloss5 = 0;
        pvalue5 = 0;
        Exvalue5 = 0;

        for K5 = 1:2*Nptlevy -1 % znegposno = -n in
Eqn
           T5 = znegpos(K1)+ znegpos(K2)+ znegpos(K3)+
znegpos(K4)+ znegpos(K5);
           Dr=1;
           if T4<= 0, Dr=0; end
           Ds =1;
       if T5 <= stoploss , Ds = 0;end
       Dt=1;
       if T5 <= T4 +stoploss, Dt=0; end% Dt relates to
trailing stoploss
       Exvalue5 = Exvalue5 + Lalphanegpos(K5)*((1-
Dr)*stoploss*(1-Dt)+Dr*(T4+stoploss)*(1-Dt)+(1-
Ds)*stoploss*Dt+Ds*T5*Dt);
       ploss5 = ploss5 + Lalphanegpos(K5)*((1-Dr)*(1-
Dt)+Dr*(1-Dt)+(1-Ds)*Dt);
       Exloss5 = Exloss5 + Lalphanegpos(K5)*((1-
Dr)*stoploss*(1-Dt)+Dr*(T4+stoploss)*(1-Dt)+(1-
Ds)*stoploss*Dt);
       pvalue5 = 1;
     end

        D =1;
        if T4 <= znegpos(znegposno) , D = 0;end
        ploss4 = ploss4 + D * Lalphanegpos(K4)*ploss5;
        pvalue4 = pvalue4 + D *
Lalphanegpos(K4)*pvalue5;
        Exloss4 = Exloss4 + D *
Lalphanegpos(K4)*Exloss5;
        Exvalue4 = Exvalue4 + D *
Lalphanegpos(K4)*Exvalue5;
      end
     D =1;
        if T3 <= znegpos(znegposno) , D = 0;end
        ploss3 = ploss3 + D * Lalphanegpos(K3)*ploss4;
        pvalue3 = pvalue3 + D *
Lalphanegpos(K3)*pvalue4;
        Exloss3 = Exloss3 + D *
Lalphanegpos(K3)*Exloss4;
```

```
        Exvalue3 = Exvalue3 + D *
Lalphanegpos(K3)*Exvalue4;
      end
        D =1;
        if T2 <= znegpos(znegposno) , D = 0;end
        ploss2 = ploss2 + D * Lalphanegpos(K2)*ploss3;
        pvalue2 = pvalue2 + D *
Lalphanegpos(K2)*pvalue3;
        Exloss2 = Exloss2 + D *
Lalphanegpos(K2)*Exloss3;
        Exvalue2 = Exvalue2 + D *
Lalphanegpos(K2)*Exvalue3;
 end
 ploss1 = ploss1+ Lalphanegpos(K1)*ploss2;
 pvalue1 = pvalue1+ Lalphanegpos(K1)*pvalue2;
 Exloss1 = Exloss1+ Lalphanegpos(K1)*Exloss2;
 Exvalue1 = Exvalue1+ Lalphanegpos(K1)*Exvalue2;
end
plossvector(t+1) = ploss1;
pvaluevector(t+1) = pvalue1;
Exlossvector(t+1) = Exloss1;
Exvaluevector(t+1) = Exvalue1;

t=6
ploss1=0;
pvalue1=0;
Exloss1=0;
Exvalue1=0;
 for K1 = znegposno +1: 2*Nptlevy -1
        T1 = znegpos(K1);
        ploss2=0;
        pvalue2 = 0;
        Exloss2=0;
        Exvalue2 = 0;
       for K2 = znegposno +1: 2*Nptlevy -1
        T2 = znegpos(K1)+ znegpos(K2);
        ploss3=0;
        pvalue3 = 0;
        Exloss3=0;
        Exvalue3 = 0;
       for K3 = znegposno +1: 2*Nptlevy -1
        T3 = znegpos(K1)+ znegpos(K2)+ znegpos(K3);
        ploss4=0;
        pvalue4 = 0;
```

```
        Exloss4=0;
        Exvalue4 = 0;
       for K4 = znegposno +1: 2*Nptlevy -1
        T4 = znegpos(K1)+ znegpos(K2)+ znegpos(K3)+
znegpos(K4);
        ploss5=0;
        pvalue5 = 0;
        Exloss5=0;
        Exvalue5 = 0;
       for K5 = znegposno +1: 2*Nptlevy -1
        T5 = znegpos(K1)+ znegpos(K2)+ znegpos(K3)+
znegpos(K4)+ znegpos(K5);
        ploss6=0;
        Exloss6 = 0;
        pvalue6 = 0;
        Exvalue6 = 0;

        for K6 = 1:2*Nptlevy -1
           T6 = znegpos(K1)+ znegpos(K2)+ znegpos(K3)+
znegpos(K4)+ znegpos(K5)+ znegpos(K6);
           Dr=1;
           if T5<= 0, Dr=0; end
           Ds =1;
       if T6 <= stoploss , Ds = 0;end
       Dt=1;
       if T6 <= T5 +stoploss, Dt=0; end% Dt relates to
trailing stoploss
       Exvalue6 = Exvalue6 + Lalphanegpos(K6)*((1-
Dr)*stoploss*(1-Dt)+Dr*(T5+stoploss)*(1-Dt)+(1-
Ds)*stoploss*Dt+Ds*T6*Dt);
       ploss6 = ploss6 + Lalphanegpos(K6)*((1-Dr)*(1-
Dt)+Dr*(1-Dt)+(1-Ds)*Dt);
       Exloss6 = Exloss6 + Lalphanegpos(K6)*((1-
Dr)*stoploss*(1-Dt)+Dr*(T5+stoploss)*(1-Dt)+(1-
Ds)*stoploss*Dt);
       pvalue6 = 1;
     end

        D =1;
        if T5 <= znegpos(znegposno) , D = 0;end
        ploss5 = ploss5 + D * Lalphanegpos(K5)*ploss6;
        pvalue5 = pvalue5 + D *
Lalphanegpos(K5)*pvalue6;
```

```
       Exloss5 = Exloss5 + D *
Lalphanegpos(K5)*Exloss6;
       Exvalue5 = Exvalue5 + D *
Lalphanegpos(K5)*Exvalue6;
     end
       D =1;
       if T4 <= znegpos(znegposno) , D = 0;end
       ploss4 = ploss4 + D * Lalphanegpos(K4)*ploss5;
       pvalue4 = pvalue4 + D *
Lalphanegpos(K4)*pvalue5;
       Exloss4 = Exloss4 + D *
Lalphanegpos(K4)*Exloss5;
       Exvalue4 = Exvalue4 + D *
Lalphanegpos(K4)*Exvalue5;
     end
    D =1;
       if T3 <= znegpos(znegposno) , D = 0;end
       ploss3 = ploss3 + D * Lalphanegpos(K3)*ploss4;
       pvalue3 = pvalue3 + D *
Lalphanegpos(K3)*pvalue4;
       Exloss3 = Exloss3 + D *
Lalphanegpos(K3)*Exloss4;
       Exvalue3 = Exvalue3 + D *
Lalphanegpos(K3)*Exvalue4;
     end
       D =1;
       if T2 <= znegpos(znegposno) , D = 0;end
       ploss2 = ploss2 + D * Lalphanegpos(K2)*ploss3;
       pvalue2 = pvalue2 + D *
Lalphanegpos(K2)*pvalue3;
       Exloss2 = Exloss2 + D *
Lalphanegpos(K2)*Exloss3;
       Exvalue2 = Exvalue2 + D *
Lalphanegpos(K2)*Exvalue3;
 end
 ploss1 = ploss1+ Lalphanegpos(K1)*ploss2;
 pvalue1 = pvalue1+ Lalphanegpos(K1)*pvalue2;
 Exloss1 = Exloss1+ Lalphanegpos(K1)*Exloss2;
 Exvalue1 = Exvalue1+ Lalphanegpos(K1)*Exvalue2;
end
plossvector(t+1) = ploss1;
pvaluevector(t+1) = pvalue1;
Exlossvector(t+1) = Exloss1;
Exvaluevector(t+1) = Exvalue1;
```

```
t=7
ploss1=0;
pvalue1=0;
Exloss1=0;
Exvalue1=0;
 for K1 = znegposno +1: 2*Nptlevy -1
       T1 = znegpos(K1);
       ploss2=0;
       pvalue2 = 0;
       Exloss2=0;
       Exvalue2 = 0;
      for K2 = znegposno +1: 2*Nptlevy -1
       T2 = znegpos(K1)+ znegpos(K2);
       ploss3=0;
       pvalue3 = 0;
       Exloss3=0;
       Exvalue3 = 0;
      for K3 = znegposno +1: 2*Nptlevy -1
       T3 = znegpos(K1)+ znegpos(K2)+ znegpos(K3);
       ploss4=0;
       pvalue4 = 0;
       Exloss4=0;
       Exvalue4 = 0;
      for K4 = znegposno +1: 2*Nptlevy -1
       T4 = znegpos(K1)+ znegpos(K2)+ znegpos(K3)+
znegpos(K4);
       ploss5=0;
       pvalue5 = 0;
       Exloss5=0;
       Exvalue5 = 0;
      for K5 = znegposno +1: 2*Nptlevy -1
       T5 = znegpos(K1)+ znegpos(K2)+ znegpos(K3)+
znegpos(K4)+ znegpos(K5);
       ploss6=0;
       pvalue6 = 0;
       Exloss6=0;
       Exvalue6 = 0;
      for K6 = znegposno +1: 2*Nptlevy -1
       T6 = znegpos(K1)+ znegpos(K2)+ znegpos(K3)+
znegpos(K4)+ znegpos(K5)+ znegpos(K6);
       ploss7=0;
       Exloss7 = 0;
       pvalue7 = 0;
       Exvalue7 = 0;
```

```
        for K7 = 1:2*Nptlevy -1
           T7 = znegpos(K1)+ znegpos(K2)+ znegpos(K3)+
znegpos(K4)+ znegpos(K5)+ znegpos(K6)+ znegpos(K7);
           Dr=1;
           if T6<= 0, Dr=0; end
           Ds =1;
       if T7 <= stoploss , Ds = 0;end
       Dt=1;
       if T7 <= T6 +stoploss, Dt=0; end% Dt relates to
trailing stoploss
       Exvalue7 = Exvalue7 + Lalphanegpos(K7)*((1-
Dr)*stoploss*(1-Dt)+Dr*(T6+stoploss)*(1-Dt)+(1-
Ds)*stoploss*Dt+Ds*T7*Dt);
       ploss7 = ploss7 + Lalphanegpos(K7)*((1-Dr)*(1-
Dt)+Dr*(1-Dt)+(1-Ds)*Dt);
       Exloss7 = Exloss7 + Lalphanegpos(K7)*((1-
Dr)*stoploss*(1-Dt)+Dr*(T6+stoploss)*(1-Dt)+(1-
Ds)*stoploss*Dt);
       pvalue7 = 1;
     end

        D=1;
        if T6 <= znegpos(znegposno) , D = 0;end
        ploss6 = ploss6 + D * Lalphanegpos(K6)*ploss7;
        pvalue6 = pvalue6 + D *
Lalphanegpos(K6)*pvalue7;
        Exloss6 = Exloss6 + D *
Lalphanegpos(K6)*Exloss7;
        Exvalue6 = Exvalue6 + D *
Lalphanegpos(K6)*Exvalue7;
      end
        D =1;
        if T5 <= znegpos(znegposno) , D = 0;end
        ploss5 = ploss5 + D * Lalphanegpos(K5)*ploss6;
        pvalue5 = pvalue5 + D *
Lalphanegpos(K5)*pvalue6;
        Exloss5 = Exloss5 + D *
Lalphanegpos(K5)*Exloss6;
        Exvalue5 = Exvalue5 + D *
Lalphanegpos(K5)*Exvalue6;
      end
        D =1;
        if T4 <= znegpos(znegposno) , D = 0;end
        ploss4 = ploss4 + D * Lalphanegpos(K4)*ploss5;
```

```
        pvalue4 = pvalue4 + D *
Lalphanegpos(K4)*pvalue5;
        Exloss4 = Exloss4 + D *
Lalphanegpos(K4)*Exloss5;
        Exvalue4 = Exvalue4 + D *
Lalphanegpos(K4)*Exvalue5;
      end
     D =1;
        if T3 <= znegpos(znegposno) , D = 0;end
        ploss3 = ploss3 + D * Lalphanegpos(K3)*ploss4;
        pvalue3 = pvalue3 + D *
Lalphanegpos(K3)*pvalue4;
        Exloss3 = Exloss3 + D *
Lalphanegpos(K3)*Exloss4;
        Exvalue3 = Exvalue3 + D *
Lalphanegpos(K3)*Exvalue4;
      end
        D =1;
        if T2 <= znegpos(znegposno) , D = 0;end
        ploss2 = ploss2 + D * Lalphanegpos(K2)*ploss3;
        pvalue2 = pvalue2 + D *
Lalphanegpos(K2)*pvalue3;
        Exloss2 = Exloss2 + D *
Lalphanegpos(K2)*Exloss3;
        Exvalue2 = Exvalue2 + D *
Lalphanegpos(K2)*Exvalue3;
 end
 ploss1 = ploss1+ Lalphanegpos(K1)*ploss2;
 pvalue1 = pvalue1+ Lalphanegpos(K1)*pvalue2;
 Exloss1 = Exloss1+ Lalphanegpos(K1)*Exloss2;
 Exvalue1 = Exvalue1+ Lalphanegpos(K1)*Exvalue2;
end
plossvector(t+1) = ploss1;
pvaluevector(t+1) = pvalue1;
Exlossvector(t+1) = Exloss1;
Exvaluevector(t+1) = Exvalue1;
t=8
ploss1=0;
pvalue1=0;
Exloss1=0;
Exvalue1=0;
 for K1 = znegposno +1: 2*Nptlevy -1
        T1 = znegpos(K1);
        ploss2=0;
```

```
        pvalue2 = 0;
        Exloss2=0;
        Exvalue2 = 0;
       for K2 = znegposno +1: 2*Nptlevy -1
        T2 = znegpos(K1)+ znegpos(K2);
        ploss3=0;
        pvalue3 = 0;
        Exloss3=0;
        Exvalue3 = 0;
       for K3 = znegposno +1: 2*Nptlevy -1
        T3 = znegpos(K1)+ znegpos(K2)+ znegpos(K3);
        ploss4=0;
        pvalue4 = 0;
        Exloss4=0;
        Exvalue4 = 0;
       for K4 = znegposno +1: 2*Nptlevy -1
        T4 = znegpos(K1)+ znegpos(K2)+ znegpos(K3)+
znegpos(K4);
        ploss5=0;
        pvalue5 = 0;
        Exloss5=0;
        Exvalue5 = 0;
       for K5 = znegposno +1: 2*Nptlevy -1
        T5 = znegpos(K1)+ znegpos(K2)+ znegpos(K3)+
znegpos(K4)+ znegpos(K5);
        ploss6=0;
        pvalue6 = 0;
        Exloss6=0;
        Exvalue6 = 0;
       for K6 = znegposno +1: 2*Nptlevy -1
        T6 = znegpos(K1)+ znegpos(K2)+ znegpos(K3)+
znegpos(K4)+ znegpos(K5)+ znegpos(K6);
        ploss7=0;
        pvalue7 = 0;
        Exloss7=0;
        Exvalue7 = 0;
       for K7 = znegposno +1: 2*Nptlevy -1
        T7 = znegpos(K1)+ znegpos(K2)+ znegpos(K3)+
znegpos(K4)+ znegpos(K5)+ znegpos(K6)+ znegpos(K7);
        ploss8=0;
        Exloss8 = 0;
        pvalue8 = 0;
        Exvalue8 = 0;
```

```
        for K8 = 1:2*Nptlevy -1
           T8 = znegpos(K1)+ znegpos(K2)+ znegpos(K3)+
znegpos(K4)+ znegpos(K5)+ znegpos(K6)+ znegpos(K7)+
znegpos(K8);
           Dr=1;
           if T7<= 0, Dr=0; end
           Ds =1;
       if T8 <= stoploss , Ds = 0;end
       Dt=1;
       if T8 <= T7 +stoploss, Dt=0; end% Dt relates to
trailing stoploss
       Exvalue8 = Exvalue8 + Lalphanegpos(K8)*((1-
Dr)*stoploss*(1-Dt)+Dr*(T7+stoploss)*(1-Dt)+(1-
Ds)*stoploss*Dt+Ds*T8*Dt);
       ploss8 = ploss8 + Lalphanegpos(K8)*((1-Dr)*(1-
Dt)+Dr*(1-Dt)+(1-Ds)*Dt);
       Exloss8 = Exloss8 + Lalphanegpos(K8)*((1-
Dr)*stoploss*(1-Dt)+Dr*(T7+stoploss)*(1-Dt)+(1-
Ds)*stoploss*Dt);
       pvalue8 = 1;
     end

        D =1;
        if T7 <= znegpos(znegposno) , D = 0;end
        ploss7 = ploss7 + D * Lalphanegpos(K7)*ploss8;
        pvalue7 = pvalue7 + D *
Lalphanegpos(K7)*pvalue8;
        Exloss7 = Exloss7 + D *
Lalphanegpos(K7)*Exloss8;
        Exvalue7 = Exvalue7 + D *
Lalphanegpos(K7)*Exvalue8;
      end
        D =1;
        if T6 <= znegpos(znegposno) , D = 0;end
        ploss6 = ploss6 + D * Lalphanegpos(K6)*ploss7;
        pvalue6 = pvalue6 + D *
Lalphanegpos(K6)*pvalue7;
        Exloss6 = Exloss6 + D *
Lalphanegpos(K6)*Exloss7;
        Exvalue6 = Exvalue6 + D *
Lalphanegpos(K6)*Exvalue7;
      end
        D =1;
        if T5 <= znegpos(znegposno) , D = 0;end
```

```
        ploss5 = ploss5 + D *
Lalphanegpos(K5)*ploss6;
        pvalue5 = pvalue5 + D *
Lalphanegpos(K5)*pvalue6;
        Exloss5 = Exloss5 + D *
Lalphanegpos(K5)*Exloss6;
        Exvalue5 = Exvalue5 + D *
Lalphanegpos(K5)*Exvalue6;
     end
       D =1;
       if T4 <= znegpos(znegposno) , D = 0;end
       ploss4 = ploss4 + D * Lalphanegpos(K4)*ploss5;
       pvalue4 = pvalue4 + D *
Lalphanegpos(K4)*pvalue5;
       Exloss4 = Exloss4 + D *
Lalphanegpos(K4)*Exloss5;
       Exvalue4 = Exvalue4 + D *
Lalphanegpos(K4)*Exvalue5;
     end
    D =1;
       if T3 <= znegpos(znegposno) , D = 0;end
       ploss3 = ploss3 + D * Lalphanegpos(K3)*ploss4;
       pvalue3 = pvalue3 + D *
Lalphanegpos(K3)*pvalue4;
       Exloss3 = Exloss3 + D *
Lalphanegpos(K3)*Exloss4;
       Exvalue3 = Exvalue3 + D *
Lalphanegpos(K3)*Exvalue4;
     end
       D =1;
       if T2 <= znegpos(znegposno) , D = 0;end
       ploss2 = ploss2 + D * Lalphanegpos(K2)*ploss3;
       pvalue2 = pvalue2 + D *
Lalphanegpos(K2)*pvalue3;
       Exloss2 = Exloss2 + D *
Lalphanegpos(K2)*Exloss3;
       Exvalue2 = Exvalue2 + D *
Lalphanegpos(K2)*Exvalue3;
 end
 ploss1 = ploss1+ Lalphanegpos(K1)*ploss2;
 pvalue1 = pvalue1+ Lalphanegpos(K1)*pvalue2;
 Exloss1 = Exloss1+ Lalphanegpos(K1)*Exloss2;
 Exvalue1 = Exvalue1+ Lalphanegpos(K1)*Exvalue2;
end
```

```
plossvector(t+1) = ploss1;
pvaluevector(t+1) = pvalue1;
Exlossvector(t+1) = Exloss1;
Exvaluevector(t+1) = Exvalue1;
t=9
ploss1=0;
pvalue1=0;
Exloss1=0;
Exvalue1=0;
 for K1 = znegposno +1: 2*Nptlevy -1
       T1 = znegpos(K1);
       ploss2=0;
       pvalue2 = 0;
       Exloss2=0;
       Exvalue2 = 0;
      for K2 = znegposno +1: 2*Nptlevy -1
       T2 = znegpos(K1)+ znegpos(K2);
       ploss3=0;
       pvalue3 = 0;
       Exloss3=0;
       Exvalue3 = 0;
      for K3 = znegposno +1: 2*Nptlevy -1
       T3 = znegpos(K1)+ znegpos(K2)+ znegpos(K3);
       ploss4=0;
       pvalue4 = 0;
       Exloss4=0;
       Exvalue4 = 0;
      for K4 = znegposno +1: 2*Nptlevy -1
       T4 = znegpos(K1)+ znegpos(K2)+ znegpos(K3)+
znegpos(K4);
       ploss5=0;
       pvalue5 = 0;
       Exloss5=0;
       Exvalue5 = 0;
      for K5 = znegposno +1: 2*Nptlevy -1
       T5 = znegpos(K1)+ znegpos(K2)+ znegpos(K3)+
znegpos(K4)+ znegpos(K5);
       ploss6=0;
       pvalue6 = 0;
       Exloss6=0;
       Exvalue6 = 0;
      for K6 = znegposno +1: 2*Nptlevy -1
       T6 = znegpos(K1)+ znegpos(K2)+ znegpos(K3)+
znegpos(K4)+ znegpos(K5)+ znegpos(K6);
```

```
        ploss7=0;
        pvalue7 = 0;
        Exloss7=0;
        Exvalue7 = 0;
       for K7 = znegposno +1: 2*Nptlevy -1
        T7 = znegpos(K1)+ znegpos(K2)+ znegpos(K3)+
znegpos(K4)+ znegpos(K5)+ znegpos(K6)+ znegpos(K7);
        ploss8=0;
        pvalue8 = 0;
        Exloss8=0;
        Exvalue8 = 0;
       for K8 = znegposno +1: 2*Nptlevy -1
        T8 = znegpos(K1)+ znegpos(K2)+ znegpos(K3)+
znegpos(K4)+ znegpos(K5)+ znegpos(K6)+ znegpos(K7)+
znegpos(K8);
        ploss9=0;
        Exloss9 = 0;
        pvalue9 = 0;
        Exvalue9 = 0;

        for K9 = 1:2*Nptlevy -1
           T9 = znegpos(K1)+ znegpos(K2)+ znegpos(K3)+
znegpos(K4)+ znegpos(K5)+ znegpos(K6)+ znegpos(K7)+
znegpos(K8)+ znegpos(K9);
           Dr=1;
           if T8<= 0, Dr=0; end
           Ds =1;
       if T9 <= stoploss , Ds = 0;end
       Dt=1;
       if T9 <= T8 +stoploss, Dt=0; end% Dt relates to
trailing stoploss
       Exvalue9 = Exvalue9 + Lalphanegpos(K9)*((1-
Dr)*stoploss*(1-Dt)+Dr*(T8+stoploss)*(1-Dt)+(1-
Ds)*stoploss*Dt+Ds*T9*Dt);
       ploss9 = ploss9 + Lalphanegpos(K9)*((1-Dr)*(1-
Dt)+Dr*(1-Dt)+(1-Ds)*Dt);
       Exloss9 = Exloss9 + Lalphanegpos(K9)*((1-
Dr)*stoploss*(1-Dt)+Dr*(T8+stoploss)*(1-Dt)+(1-
Ds)*stoploss*Dt);
       pvalue9 = 1;
     end

        D =1;
        if T8 <= znegpos(znegposno) , D = 0;end
```

```
        ploss8 = ploss8 + D * Lalphanegpos(K8)*ploss9;
        pvalue8 = pvalue8 + D *
Lalphanegpos(K8)*pvalue9;
        Exloss8 = Exloss8 + D *
Lalphanegpos(K8)*Exloss9;
        Exvalue8 = Exvalue8 + D *
Lalphanegpos(K8)*Exvalue9;
      end
        D =1;
        if T7 <= znegpos(znegposno) , D = 0;end
        ploss7 = ploss7 + D * Lalphanegpos(K7)*ploss8;
        pvalue7 = pvalue7 + D *
Lalphanegpos(K7)*pvalue8;
        Exloss7 = Exloss7 + D *
Lalphanegpos(K7)*Exloss8;
        Exvalue7 = Exvalue7 + D *
Lalphanegpos(K7)*Exvalue8;
      end
        D =1;
        if T6 <= znegpos(znegposno) , D = 0;end
        ploss6 = ploss6 + D * Lalphanegpos(K6)*ploss7;
        pvalue6 = pvalue6 + D *
Lalphanegpos(K6)*pvalue7;
        Exloss6 = Exloss6 + D *
Lalphanegpos(K6)*Exloss7;
        Exvalue6 = Exvalue6 + D *
Lalphanegpos(K6)*Exvalue7;
      end
        D =1;
        if T5 <= znegpos(znegposno) , D = 0;end
        ploss5 = ploss5 + D * Lalphanegpos(K5)*ploss6;
        pvalue5 = pvalue5 + D *
Lalphanegpos(K5)*pvalue6;
        Exloss5 = Exloss5 + D *
Lalphanegpos(K5)*Exloss6;
        Exvalue5 = Exvalue5 + D *
Lalphanegpos(K5)*Exvalue6;
      end
        D =1;
        if T4 <= znegpos(znegposno) , D = 0;end
        ploss4 = ploss4 + D * Lalphanegpos(K4)*ploss5;
        pvalue4 = pvalue4 + D *
Lalphanegpos(K4)*pvalue5;
```

```
       Exloss4 = Exloss4 + D *
Lalphanegpos(K4)*Exloss5;
       Exvalue4 = Exvalue4 + D *
Lalphanegpos(K4)*Exvalue5;
     end
    D =1;
       if T3 <= znegpos(znegposno) , D = 0;end
       ploss3 = ploss3 + D * Lalphanegpos(K3)*ploss4;
       pvalue3 = pvalue3 + D *
Lalphanegpos(K3)*pvalue4;
       Exloss3 = Exloss3 + D *
Lalphanegpos(K3)*Exloss4;
       Exvalue3 = Exvalue3 + D *
Lalphanegpos(K3)*Exvalue4;
     end
       D =1;
       if T2 <= znegpos(znegposno) , D = 0;end
       ploss2 = ploss2 + D * Lalphanegpos(K2)*ploss3;
       pvalue2 = pvalue2 + D *
Lalphanegpos(K2)*pvalue3;
       Exloss2 = Exloss2 + D *
Lalphanegpos(K2)*Exloss3;
       Exvalue2 = Exvalue2 + D *
Lalphanegpos(K2)*Exvalue3;
 end
 ploss1 = ploss1+ Lalphanegpos(K1)*ploss2;
 pvalue1 = pvalue1+ Lalphanegpos(K1)*pvalue2;
 Exloss1 = Exloss1+ Lalphanegpos(K1)*Exloss2;
 Exvalue1 = Exvalue1+ Lalphanegpos(K1)*Exvalue2;
end
plossvector(t+1) = ploss1;
pvaluevector(t+1) = pvalue1;
Exlossvector(t+1) = Exloss1;
Exvaluevector(t+1) = Exvalue1;
t=10
ploss1=0;
pvalue1=0;
Exloss1=0;
Exvalue1=0;
 for K1 = znegposno +1: 2*Nptlevy -1
       T1 = znegpos(K1);
       ploss2=0;
       pvalue2 = 0;
       Exloss2=0;
```

```
       Exvalue2 = 0;
      for K2 = znegposno +1: 2*Nptlevy -1
       T2 = znegpos(K1)+ znegpos(K2);
       ploss3=0;
       pvalue3 = 0;
       Exloss3=0;
       Exvalue3 = 0;
      for K3 = znegposno +1: 2*Nptlevy -1
       T3 = znegpos(K1)+ znegpos(K2)+ znegpos(K3);
       ploss4=0;
       pvalue4 = 0;
       Exloss4=0;
       Exvalue4 = 0;
      for K4 = znegposno +1: 2*Nptlevy -1
       T4 = znegpos(K1)+ znegpos(K2)+ znegpos(K3)+
znegpos(K4);
       ploss5=0;
       pvalue5 = 0;
       Exloss5=0;
       Exvalue5 = 0;
      for K5 = znegposno +1: 2*Nptlevy -1
       T5 = znegpos(K1)+ znegpos(K2)+ znegpos(K3)+
znegpos(K4)+ znegpos(K5);
       ploss6=0;
       pvalue6 = 0;
       Exloss6=0;
       Exvalue6 = 0;
      for K6 = znegposno +1: 2*Nptlevy -1
       T6 = znegpos(K1)+ znegpos(K2)+ znegpos(K3)+
znegpos(K4)+ znegpos(K5)+ znegpos(K6);
       ploss7=0;
       pvalue7 = 0;
       Exloss7=0;
       Exvalue7 = 0;
      for K7 = znegposno +1: 2*Nptlevy -1
       T7 = znegpos(K1)+ znegpos(K2)+ znegpos(K3)+
znegpos(K4)+ znegpos(K5)+ znegpos(K6)+ znegpos(K7);
       ploss8=0;
       pvalue8 = 0;
       Exloss8=0;
       Exvalue8 = 0;
       for K8 = znegposno +1: 2*Nptlevy -1
```

```
       T8 = znegpos(K1)+ znegpos(K2)+ znegpos(K3)+
znegpos(K4)+ znegpos(K5)+ znegpos(K6)+ znegpos(K7)+
znegpos(K8);
       ploss9=0;
       pvalue9 = 0;
       Exloss9=0;
       Exvalue9 = 0;
      for K9 = znegposno +1: 2*Nptlevy -1
       T9 = znegpos(K1)+ znegpos(K2)+ znegpos(K3)+
znegpos(K4)+ znegpos(K5)+ znegpos(K6)+ znegpos(K7)+
znegpos(K8)+ znegpos(K9);
       ploss10=0;
       Exloss10 = 0;
       pvalue10 = 0;
       Exvalue10 = 0;

       for K10 = 1:2*Nptlevy -1
          T10 = znegpos(K1)+ znegpos(K2)+
znegpos(K3)+ znegpos(K4)+ znegpos(K5)+ znegpos(K6)+
znegpos(K7)+ znegpos(K8)+ znegpos(K9)+ znegpos(K10);
          Dr=1;
          if T9<= 0, Dr=0; end
          Ds =1;
      if T10 <= stoploss , Ds = 0;end
      Dt=1;
      if T10 <= T9 +stoploss, Dt=0; end% Dt relates
to trailing stoploss
      Exvalue10 = Exvalue10 + Lalphanegpos(K10)*((1-
Dr)*stoploss*(1-Dt)+Dr*(T9+stoploss)*(1-Dt)+(1-
Ds)*stoploss*Dt+Ds*T10*Dt);
      ploss10 = ploss10 + Lalphanegpos(K10)*((1-
Dr)*(1-Dt)+Dr*(1-Dt)+(1-Ds)*Dt);
      Exloss10 = Exloss10 + Lalphanegpos(K10)*((1-
Dr)*stoploss*(1-Dt)+Dr*(T9+stoploss)*(1-Dt)+(1-
Ds)*stoploss*Dt);
      pvalue10 = 1;
    end

       D =1;
       if T9 <= znegpos(znegposno) , D = 0;end
       ploss9 = ploss9 + D *
Lalphanegpos(K9)*ploss10;
       pvalue9 = pvalue9 + D *
Lalphanegpos(K9)*pvalue10;
```

```
        Exloss9 = Exloss9 + D *
Lalphanegpos(K9)*Exloss10;
        Exvalue9 = Exvalue9 + D *
Lalphanegpos(K9)*Exvalue10;
      end
        D =1;
        if T8 <= znegpos(znegposno) , D = 0;end
        ploss8 = ploss8 + D * Lalphanegpos(K8)*ploss9;
        pvalue8 = pvalue8 + D *
Lalphanegpos(K8)*pvalue9;
        Exloss8 = Exloss8 + D *
Lalphanegpos(K8)*Exloss9;
        Exvalue8 = Exvalue8 + D *
Lalphanegpos(K8)*Exvalue9;
      end
        D =1;
        if T7 <= znegpos(znegposno) , D = 0;end
        ploss7 = ploss7 + D * Lalphanegpos(K7)*ploss8;
        pvalue7 = pvalue7 + D *
Lalphanegpos(K7)*pvalue8;
        Exloss7 = Exloss7 + D *
Lalphanegpos(K7)*Exloss8;
        Exvalue7 = Exvalue7 + D *
Lalphanegpos(K7)*Exvalue8;
      end
        D =1;
        if T6 <= znegpos(znegposno) , D = 0;end
        ploss6 = ploss6 + D * Lalphanegpos(K6)*ploss7;
        pvalue6 = pvalue6 + D *
Lalphanegpos(K6)*pvalue7;
        Exloss6 = Exloss6 + D *
Lalphanegpos(K6)*Exloss7;
        Exvalue6 = Exvalue6 + D *
Lalphanegpos(K6)*Exvalue7;
      end
        D =1;
        if T5 <= znegpos(znegposno) , D = 0;end
        ploss5 = ploss5 + D * Lalphanegpos(K5)*ploss6;
        pvalue5 = pvalue5 + D *
Lalphanegpos(K5)*pvalue6;
        Exloss5 = Exloss5 + D *
Lalphanegpos(K5)*Exloss6;
        Exvalue5 = Exvalue5 + D *
Lalphanegpos(K5)*Exvalue6;
```

```
      end
        D =1;
        if T4 <= znegpos(znegposno) , D = 0;end
        ploss4 = ploss4 + D * Lalphanegpos(K4)*ploss5;
        pvalue4 = pvalue4 + D *
Lalphanegpos(K4)*pvalue5;
        Exloss4 = Exloss4 + D *
Lalphanegpos(K4)*Exloss5;
        Exvalue4 = Exvalue4 + D *
Lalphanegpos(K4)*Exvalue5;
      end
     D =1;
        if T3 <= znegpos(znegposno) , D = 0;end
        ploss3 = ploss3 + D * Lalphanegpos(K3)*ploss4;
        pvalue3 = pvalue3 + D *
Lalphanegpos(K3)*pvalue4;
        Exloss3 = Exloss3 + D *
Lalphanegpos(K3)*Exloss4;
        Exvalue3 = Exvalue3 + D *
Lalphanegpos(K3)*Exvalue4;
      end
        D =1;
        if T2 <= znegpos(znegposno) , D = 0;end
        ploss2 = ploss2 + D * Lalphanegpos(K2)*ploss3;
        pvalue2 = pvalue2 + D *
Lalphanegpos(K2)*pvalue3;
        Exloss2 = Exloss2 + D *
Lalphanegpos(K2)*Exloss3;
        Exvalue2 = Exvalue2 + D *
Lalphanegpos(K2)*Exvalue3;
 end
 ploss1 = ploss1+ Lalphanegpos(K1)*ploss2;
 pvalue1 = pvalue1+ Lalphanegpos(K1)*pvalue2;
 Exloss1 = Exloss1+ Lalphanegpos(K1)*Exloss2;
 Exvalue1 = Exvalue1+ Lalphanegpos(K1)*Exvalue2;
end
plossvector(t+1) = ploss1;
pvaluevector(t+1) = pvalue1;
Exlossvector(t+1) = Exloss1;
Exvaluevector(t+1) = Exvalue1;


for I =1:t+1
   time(I) = I-1;
```

```
end
figure(4)
plot(time,Exvaluevector,'k*-')
xlabel(' t ')
ylabel('Expected value at time t,   with trailing
stop-loss')
title('S & P 500')

Totalp(2)=1;% As Lalphanegpos is normalized.  The
total probability is 1 at t=1
Totalex(2)=Exvaluevector(2);
plosssum=0;
Explosssum=0;
for I=3:t+1
   plosssum = plosssum + plossvector(I-1);% add up
prob of stopping out in previous days, 35M142
   Exlosssum = Exlosssum+Exlossvector(I-1);% add up
expected values of stopping out for previous days
   Totalp(I)=plosssum+pvaluevector(I);% Totalp is the
sum of the probability of previous days losses and
last day's cshing out
   Totalex(I)=Exlosssum+Exvaluevector(I);%Totalex is
the expected value of previous days losses and last
day's gain
end
   Expectedt(2)=1;%Expectedt is the average number of
t units of the trade when the trader chooses to cash
out on the t th unit
for I=3:t+1
   Sum=0;
   for J=2:I-1
      Sum =Sum + plossvector(J)*(J-1);
   end
   Expectedt(I)= Sum +pvaluevector(I)*(I-1);
end
for I=2:t+1
   TotalexDt(I)=Totalex(I)/Expectedt(I);
end
figure(5)
plot(time(2:t+1),Exlossvector(2:t+1),'k*-')% last
point of Exlossvector actually not used in
calculation.  It is calculated for plotting purpose
only.
xlabel(' t ')
```

```
ylabel('Expected loss at time t, with trailing stop
loss ')
title('S & P 500')
figure(6)
plot(time(2:t+1),Totalex(2:t+1),'k*-')
xlabel(' t ')
ylabel('Total expected value, with trailing stop
loss')
title('S & P 500')
figure(7)
plot(time(2:t+1),plossvector(2:t+1),'k*-')
xlabel(' t ')
ylabel('plossvector')
title('S & P 500')
figure(8)
plot(time(2:t+1),pvaluevector(2:t+1),'k*-')
xlabel(' t ')
ylabel('pvaluevector, with trailing stop loss')
title('S & P 500')
figure(9)
plot(time(2:t+1),Totalp(2:t+1),'k*-')
xlabel(' t ')
ylabel('Totalp, with trailing stop loss')
title('S & P 500')
figure(10)
plot(time(2:t+1),Expectedt(2:t+1),'k*-')
xlabel(' t ')
ylabel('Average time, with trailing stop loss')
title('S & P 500')
figure(11)
plot(time(2:t+1),TotalexDt(2:t+1),'k*-')
xlabel(' t ')
ylabel('Total expected value/Average time , with
trailing stop-loss')
title('S & P 500')
```

A5.2 Fixed Stop-Loss Program

```
%Fixedstoploss
%
clear
t=10;
for I=1:t+1
   %day is only an example of one time unit
   plossvector(I)=0;% probability of that day's
stopping out by initial stoploss or trailing stoploss
   Exlossvector(I)=0;% expected value of being
stopped out
   pvaluevector(I)=0;% probability of last day' cash
out
   Totalp(I)=0;% Total probability of previous days'
stops and last day cash out
   Totalex(I)=0;
   Expectedt(I)=0;
end
deltat=1;
interval = .05;%for calculating the integral
endpoint = 500;% for calculating the integral
q = 0: interval: endpoint;
zinterval = 0.025;  %
zendpoint = 0.075;%0.075
z = 0: zinterval: zendpoint;
Nptlevy = round(zendpoint/zinterval) +1 ;
for J=1:Nptlevy
   for I=1:endpoint/interval +1
   y(I)= (1/pi)*exp(-
0.00375*deltat*q(I)^1.4)*cos(q(I)*z(J));% calculate
Levy distribution
   end
   Lalpha(J)=trapz(q,y); %using trapezoidal approx to
calculate the integral
end
figure(1)
plot(z,Lalpha,'k*-')
xlabel('Z,   Delta t = 1')
ylabel('Levy distribution')
title('')
% To assign numbers to the symmetrical Levy function
for J=1: 2*Nptlevy -1
   znegpos(J)=0;
```

```
   Lalphanegpos(J)=0;
end
   for J = 1:Nptlevy - 1
   znegpos(J) = -z(Nptlevy+1 - J);
   Lalphanegpos(J) = Lalpha(Nptlevy+1 - J);
end
for J= 1: Nptlevy
   znegpos(J + Nptlevy -1) = z(J);
   Lalphanegpos(J + Nptlevy -1) = Lalpha(J);
end
figure(2)
plot(znegpos,Lalphanegpos,'k*-')
xlabel('Z,   Delta t = 1')
ylabel('Levy distribution   ')
title('')
Sumlevy=0;% Normalize the Levy dist.
for I=1:2*Nptlevy -1
   Sumlevy = Sumlevy + Lalphanegpos(I);
end
for I=1:2*Nptlevy -1
   Lalphanegpos(I) = Lalphanegpos(I)/Sumlevy;
end
figure(3)
plot(znegpos,Lalphanegpos,'k*-')
xlabel('Z,   Delta t = 1')
ylabel('Normalized Levy distribution')
title('Normalized prob dist')
%Nptlevy = 2;% for testing only
%znegpos=[-1 0 1 ];% for testing only
%Lalphanegpos = [ 0.25  0.5  0.25];% for testing only
%Nptlevy = 3;% for testing only
%znegpos=[-2 -1 0 1 2];% for testing only
%Lalphanegpos = [0.1  0.2  0.4  0.2  0.1];% for
testing only
S= -1;%                  change stoploss   S = 0, -1,
-2,...   ***********
znegposno = Nptlevy + S;% this is the number in the
vector where stoploss is set
% Buy at t = 0, find the expectation value at furture
t, stoploss is fixed
stoplossvalue=znegpos(znegposno);% stoploss value is
fixed thru out
Exvaluevector(1) = 0; % at time t=0,  buy --
therefore no gain or loss
```

```
t=1% t = 1 time unit
ploss1=0;
pvalue1 = 0;
Exloss1=0;
Exvalue1 = 0;
    for K1 = 1:2*Nptlevy -1 % znegposno = -n in
35M107
      T1 = znegpos(K1);
      D =1;
      if T1 <= znegpos(znegposno) , D = 0;end
      ploss1 = ploss1 + Lalphanegpos(K1)*(1-D);
      pvalue1 = pvalue1 + Lalphanegpos(K1)*((1-
D)+D);%pvalue1 will add up to 1
      Exloss1 = Exloss1 +
Lalphanegpos(K1)*stoplossvalue*(1-D);
      Exvalue1 = Exvalue1 +
Lalphanegpos(K1)*(stoplossvalue*(1-D)+T1*D);
    end
plossvector(t+1) = ploss1;
pvaluevector(t+1) = pvalue1;
Exlossvector(t+1) = Exloss1;
Exvaluevector(t+1) = Exvalue1;
t=2
ploss1=0;
pvalue1 = 0;
Exloss1=0;
Exvalue1=0;
for K1 = 1: 2*Nptlevy -1
       T1 = znegpos(K1);
       ploss2 = 0;
       pvalue2 = 0;
       Exloss2 = 0;
       Exvalue2 = 0;
    for K2 = 1:2*Nptlevy -1
      T2 = znegpos(K1)+ znegpos(K2);
      D =1;
      if T2 <= znegpos(znegposno) , D = 0;end
      ploss2 = ploss2 + Lalphanegpos(K2)*(1-D);
      pvalue2 =pvalue2 + Lalphanegpos(K2)*((1-D)+D);
      Exloss2 = Exloss2 +
Lalphanegpos(K2)*stoplossvalue*(1-D);
      Exvalue2 = Exvalue2 +
Lalphanegpos(K2)*(stoplossvalue*(1-D)+T2*D);
    end
```

```
    D=1;
    if T1 <= znegpos(znegposno) , D = 0;end
    ploss1 = ploss1 + D * Lalphanegpos(K1)*ploss2;
    pvalue1 = pvalue1 + D * Lalphanegpos(K1)*pvalue2;
    Exloss1 = Exloss1 + D * Lalphanegpos(K1)*Exloss2;
    Exvalue1 = Exvalue1 + D *
Lalphanegpos(K1)*Exvalue2;
 end
 plossvector(t+1) = ploss1;
 pvaluevector(t+1) = pvalue1;
 Exlossvector(t+1) = Exloss1;
 Exvaluevector(t+1) = Exvalue1;
 t=3
 ploss1=0;
 pvalue1 = 0;
 Exloss1=0;
 Exvalue1=0;
for K1 = 1: 2*Nptlevy -1
       T1 = znegpos(K1);
       ploss2=0;
       pvalue2 = 0;
       Exloss2=0;
       Exvalue2 = 0;
 for K2 = 1: 2*Nptlevy -1
       T2 = znegpos(K1)+znegpos(K2);
       ploss3=0;
       pvalue3 = 0;
       Exloss3=0;
       Exvalue3 = 0;
   for K3 = 1:2*Nptlevy -1
      T3 = znegpos(K1)+ znegpos(K2)+ znegpos(K3);
      D =1;
      if T3 <= znegpos(znegposno) , D = 0;end
      ploss3 = ploss3 + Lalphanegpos(K3)*(1-D);
      pvalue3 = pvalue3 + Lalphanegpos(K3)*((1-D)+D);
      Exloss3 = Exloss3 +
Lalphanegpos(K3)*stoplossvalue*(1-D);
      Exvalue3 = Exvalue3 +
Lalphanegpos(K3)*(stoplossvalue*(1-D)+T3*D);
    end
    D=1;
    if T2 <= znegpos(znegposno) , D = 0;end
    ploss2 = ploss2 + D * Lalphanegpos(K2)*ploss3;
    pvalue2 = pvalue2 + D * Lalphanegpos(K2)*pvalue3;
```

```
    Exloss2 = Exloss2 + D * Lalphanegpos(K2)*Exloss3;
    Exvalue2 = Exvalue2 + D *
Lalphanegpos(K2)*Exvalue3;
 end
    D=1;
    if T1 <= znegpos(znegposno) , D = 0;end
    ploss1 = ploss1 + D * Lalphanegpos(K1)*ploss2;
    pvalue1 = pvalue1 + D * Lalphanegpos(K1)*pvalue2;
    Exloss1 = Exloss1 + D * Lalphanegpos(K1)*Exloss2;
    Exvalue1 = Exvalue1 + D *
Lalphanegpos(K1)*Exvalue2;
 end
 plossvector(t+1) = ploss1;
 pvaluevector(t+1) = pvalue1;
 Exlossvector(t+1) = Exloss1;
 Exvaluevector(t+1) = Exvalue1;
t=4
ploss1=0;
pvalue1=0;
Exloss1=0;
Exvalue1=0;
for K1 = 1: 2*Nptlevy -1
       T1 = znegpos(K1);
       ploss2=0;
       pvalue2 = 0;
       Exloss2=0;
       Exvalue2 = 0;
 for K2 = 1: 2*Nptlevy -1
       T2 = znegpos(K1)+znegpos(K2);
       ploss3=0;
       pvalue3 = 0;
       Exloss3=0;
       Exvalue3 = 0;
    for K3 = 1: 2*Nptlevy -1
       T3 = znegpos(K1)+znegpos(K2)+ znegpos(K3);
       ploss4=0;
       pvalue4 = 0;
       Exloss4=0;
       Exvalue4 = 0;
    for K4 = 1:2*Nptlevy -1
      T4 = znegpos(K1)+ znegpos(K2)+ znegpos(K3)+
znegpos(K4);
      D =1;
      if T4 <= znegpos(znegposno) , D = 0;end
```

```
      ploss4 = ploss4 + Lalphanegpos(K4)*(1-D);
      pvalue4 = pvalue4 + Lalphanegpos(K4)*((1-
D)+D);% pvalue4 should add up to be 1
      Exloss4 = Exloss4 +
Lalphanegpos(K4)*stoplossvalue*(1-D);
      Exvalue4 = Exvalue4 +
Lalphanegpos(K4)*(stoplossvalue*(1-D)+T4*D);
    end
    D=1;
    if T3 <= znegpos(znegposno) , D = 0;end
    ploss3 = ploss3 + D * Lalphanegpos(K3)*ploss4;
    pvalue3 = pvalue3 + D * Lalphanegpos(K3)*pvalue4;
    Exloss3 = Exloss3 + D * Lalphanegpos(K3)*Exloss4;
    Exvalue3 = Exvalue3 + D *
Lalphanegpos(K3)*Exvalue4;
 end
    D=1;
    if T2 <= znegpos(znegposno) , D = 0;end
    ploss2 = ploss2 + D * Lalphanegpos(K2)*ploss3;
    pvalue2 = pvalue2 + D * Lalphanegpos(K2)*pvalue3;
    Exloss2 = Exloss2 + D * Lalphanegpos(K2)*Exloss3;
    Exvalue2 = Exvalue2 + D *
Lalphanegpos(K2)*Exvalue3;
 end
    D=1;
    if T1 <= znegpos(znegposno) , D = 0;end
    ploss1 = ploss1 + D * Lalphanegpos(K1)*ploss2;
    pvalue1 = pvalue1 + D * Lalphanegpos(K1)*pvalue2;
    Exloss1 = Exloss1 + D * Lalphanegpos(K1)*Exloss2;
    Exvalue1 = Exvalue1 + D *
Lalphanegpos(K1)*Exvalue2;
 end
 plossvector(t+1) = ploss1;
 pvaluevector(t+1) = pvalue1;
 Exlossvector(t+1) = Exloss1;
 Exvaluevector(t+1) = Exvalue1;
t=5
ploss1=0;
pvalue1=0;
Exloss1=0;
Exvalue1=0;
for K1 = 1: 2*Nptlevy -1
       T1 = znegpos(K1);
       ploss2=0;
```

```
        pvalue2 = 0;
        Exloss2=0;
        Exvalue2 = 0;
 for K2 = 1: 2*Nptlevy -1
        T2 = znegpos(K1)+znegpos(K2);
        ploss3=0;
        pvalue3 = 0;
        Exloss3=0;
        Exvalue3 = 0;
     for K3 = 1: 2*Nptlevy -1
        T3 = znegpos(K1)+znegpos(K2)+ znegpos(K3);
        ploss4=0;
        pvalue4 = 0;
        Exloss4=0;
        Exvalue4 = 0;
     for K4 = 1: 2*Nptlevy -1
        T4 = znegpos(K1)+znegpos(K2)+ znegpos(K3)+
znegpos(K4);
        ploss5=0;
        pvalue5 = 0;
        Exloss5=0;
        Exvalue5 = 0;
     for K5 = 1:2*Nptlevy -1
       T5 = znegpos(K1)+ znegpos(K2)+ znegpos(K3)+
znegpos(K4)+ znegpos(K5);
       D =1;
       if T5 <= znegpos(znegposno) , D = 0;end
       ploss5 = ploss5 + Lalphanegpos(K5)*(1-D);
       pvalue5 = pvalue5 + Lalphanegpos(K5)*((1-D)+D);
       Exloss5 = Exloss5 +
Lalphanegpos(K5)*stoplossvalue*(1-D);
       Exvalue5 = Exvalue5 +
Lalphanegpos(K5)*(stoplossvalue*(1-D)+T5*D);
     end
      D=1;
     if T4 <= znegpos(znegposno) , D = 0;end
     ploss4 = ploss4 + D * Lalphanegpos(K4)*ploss5;
     pvalue4 = pvalue4 + D * Lalphanegpos(K4)*pvalue5;
     Exloss4 = Exloss4 + D * Lalphanegpos(K4)*Exloss5;
     Exvalue4 = Exvalue4 + D *
Lalphanegpos(K4)*Exvalue5;
  end
     D=1;
     if T3 <= znegpos(znegposno) , D = 0;end
```

```
    ploss3 = ploss3 + D * Lalphanegpos(K3)*ploss4;
    pvalue3 = pvalue3 + D * Lalphanegpos(K3)*pvalue4;
    Exloss3 = Exloss3 + D * Lalphanegpos(K3)*Exloss4;
    Exvalue3 = Exvalue3 + D *
Lalphanegpos(K3)*Exvalue4;
 end
    D=1;
    if T2 <= znegpos(znegposno) , D = 0;end
    ploss2 = ploss2 + D * Lalphanegpos(K2)*ploss3;
    pvalue2 = pvalue2 + D * Lalphanegpos(K2)*pvalue3;
    Exloss2 = Exloss2 + D * Lalphanegpos(K2)*Exloss3;
    Exvalue2 = Exvalue2 + D *
Lalphanegpos(K2)*Exvalue3;
 end
    D=1;
    if T1 <= znegpos(znegposno) , D = 0;end
    ploss1 = ploss1 + D * Lalphanegpos(K1)*ploss2;
    pvalue1 = pvalue1 + D * Lalphanegpos(K1)*pvalue2;
    Exloss1 = Exloss1 + D * Lalphanegpos(K1)*Exloss2;
    Exvalue1 = Exvalue1 + D *
Lalphanegpos(K1)*Exvalue2;
 end
 plossvector(t+1) = ploss1;
 pvaluevector(t+1) = pvalue1;
 Exlossvector(t+1) = Exloss1;
 Exvaluevector(t+1) = Exvalue1;
t=6
ploss1=0;
pvalue1=0;
Exloss1=0;
Exvalue1=0;
for K1 = 1: 2*Nptlevy -1
       T1 = znegpos(K1);
       ploss2=0;
       pvalue2 = 0;
       Exloss2=0;
       Exvalue2 = 0;
 for K2 = 1: 2*Nptlevy -1
       T2 = znegpos(K1)+znegpos(K2);
       ploss3=0;
       pvalue3 = 0;
       Exloss3=0;
       Exvalue3 = 0;
    for K3 = 1: 2*Nptlevy -1
```

```
        T3 = znegpos(K1)+znegpos(K2)+ znegpos(K3);
        ploss4=0;
        pvalue4 = 0;
        Exloss4=0;
        Exvalue4 = 0;
     for K4 = 1: 2*Nptlevy -1
        T4 = znegpos(K1)+znegpos(K2)+ znegpos(K3)+
znegpos(K4);
        ploss5=0;
        pvalue5 = 0;
        Exloss5=0;
        Exvalue5 = 0;
     for K5 = 1: 2*Nptlevy -1
        T5 = znegpos(K1)+znegpos(K2)+ znegpos(K3)+
znegpos(K4)+ znegpos(K5);
        ploss6=0;
        pvalue6 = 0;
        Exloss6=0;
        Exvalue6 = 0;
     for K6 = 1:2*Nptlevy -1
       T6 = znegpos(K1)+ znegpos(K2)+ znegpos(K3)+
znegpos(K4)+ znegpos(K5)+ znegpos(K6);
       D =1;
       if T6 <= znegpos(znegposno) , D = 0;end
       ploss6 = ploss6 + Lalphanegpos(K6)*(1-D);
       pvalue6 = pvalue6 + Lalphanegpos(K6)*((1-D)+D);
       Exloss6 = Exloss6 +
Lalphanegpos(K6)*stoplossvalue*(1-D);
       Exvalue6 = Exvalue6 +
Lalphanegpos(K6)*(stoplossvalue*(1-D)+T6*D);
     end
      D=1;
     if T5 <= znegpos(znegposno) , D = 0;end
     ploss5 = ploss5 + D * Lalphanegpos(K5)*ploss6;
     pvalue5 = pvalue5 + D * Lalphanegpos(K5)*pvalue6;
     Exloss5 = Exloss5 + D * Lalphanegpos(K5)*Exloss6;
     Exvalue5 = Exvalue5 + D *
Lalphanegpos(K5)*Exvalue6;
 end
     D=1;
     if T4 <= znegpos(znegposno) , D = 0;end
     ploss4 = ploss4 + D * Lalphanegpos(K4)*ploss5;
     pvalue4 = pvalue4 + D * Lalphanegpos(K4)*pvalue5;
     Exloss4 = Exloss4 + D * Lalphanegpos(K4)*Exloss5;
```

```
    Exvalue4 = Exvalue4 + D *
Lalphanegpos(K4)*Exvalue5;
 end
    D=1;
    if T3 <= znegpos(znegposno) , D = 0;end
    ploss3 = ploss3 + D * Lalphanegpos(K3)*ploss4;
    pvalue3 = pvalue3 + D * Lalphanegpos(K3)*pvalue4;
    Exloss3 = Exloss3 + D * Lalphanegpos(K3)*Exloss4;
    Exvalue3 = Exvalue3 + D *
Lalphanegpos(K3)*Exvalue4;
 end
    D=1;
    if T2 <= znegpos(znegposno) , D = 0;end
    ploss2 = ploss2 + D * Lalphanegpos(K2)*ploss3;
    pvalue2 = pvalue2 + D * Lalphanegpos(K2)*pvalue3;
    Exloss2 = Exloss2 + D * Lalphanegpos(K2)*Exloss3;
    Exvalue2 = Exvalue2 + D *
Lalphanegpos(K2)*Exvalue3;
 end
    D=1;
    if T1 <= znegpos(znegposno) , D = 0;end
    ploss1 = ploss1 + D * Lalphanegpos(K1)*ploss2;
    pvalue1 = pvalue1 + D * Lalphanegpos(K1)*pvalue2;
    Exloss1 = Exloss1 + D * Lalphanegpos(K1)*Exloss2;
    Exvalue1 = Exvalue1 + D *
Lalphanegpos(K1)*Exvalue2;
 end
 plossvector(t+1) = ploss1;
 pvaluevector(t+1) = pvalue1;
 Exlossvector(t+1) = Exloss1;
 Exvaluevector(t+1) = Exvalue1;
t=7
ploss1=0;
pvalue1=0;
Exloss1=0;
Exvalue1=0;
for K1 = 1: 2*Nptlevy -1
       T1 = znegpos(K1);
       ploss2=0;
       pvalue2 = 0;
       Exloss2=0;
       Exvalue2 = 0;
 for K2 = 1: 2*Nptlevy -1
       T2 = znegpos(K1)+znegpos(K2);
```

```
        ploss3=0;
        pvalue3 = 0;
        Exloss3=0;
        Exvalue3 = 0;
    for K3 = 1: 2*Nptlevy -1
        T3 = znegpos(K1)+znegpos(K2)+ znegpos(K3);
        ploss4=0;
        pvalue4 = 0;
        Exloss4=0;
        Exvalue4 = 0;
    for K4 = 1: 2*Nptlevy -1
        T4 = znegpos(K1)+znegpos(K2)+ znegpos(K3)+
znegpos(K4);
        ploss5=0;
        pvalue5 = 0;
        Exloss5=0;
        Exvalue5 = 0;
    for K5 = 1: 2*Nptlevy -1
        T5 = znegpos(K1)+znegpos(K2)+ znegpos(K3)+
znegpos(K4)+ znegpos(K5);
        ploss6=0;
        pvalue6 = 0;
        Exloss6=0;
        Exvalue6 = 0;
    for K6 = 1: 2*Nptlevy -1
        T6 = znegpos(K1)+znegpos(K2)+ znegpos(K3)+
znegpos(K4)+ znegpos(K5)+ znegpos(K6);
        ploss7=0;
        pvalue7 = 0;
        Exloss7=0;
        Exvalue7 = 0;
    for K7 = 1:2*Nptlevy -1
       T7 = znegpos(K1)+ znegpos(K2)+ znegpos(K3)+
znegpos(K4)+ znegpos(K5)+ znegpos(K6)+ znegpos(K7);
       D =1;
       if T7 <= znegpos(znegposno) , D = 0;end
       ploss7 = ploss7 + Lalphanegpos(K7)*(1-D);
       pvalue7 = pvalue7 + Lalphanegpos(K7)*((1-D)+D);
       Exloss7 = Exloss7 +
Lalphanegpos(K7)*stoplossvalue*(1-D);
       Exvalue7 = Exvalue7 +
Lalphanegpos(K7)*(stoplossvalue*(1-D)+T7*D);
    end
     D=1;
```

```
    if T6 <= znegpos(znegposno) , D = 0;end
    ploss6 = ploss6 + D * Lalphanegpos(K6)*ploss7;
    pvalue6 = pvalue6 + D * Lalphanegpos(K6)*pvalue7;
    Exloss6 = Exloss6 + D * Lalphanegpos(K6)*Exloss7;
    Exvalue6 = Exvalue6 + D *
Lalphanegpos(K6)*Exvalue7;
 end
    D=1;
    if T5 <= znegpos(znegposno) , D = 0;end
    ploss5 = ploss5 + D * Lalphanegpos(K5)*ploss6;
    pvalue5 = pvalue5 + D * Lalphanegpos(K5)*pvalue6;
    Exloss5 = Exloss5 + D * Lalphanegpos(K5)*Exloss6;
    Exvalue5 = Exvalue5 + D *
Lalphanegpos(K5)*Exvalue6;
 end
    D=1;
    if T4 <= znegpos(znegposno) , D = 0;end
    ploss4 = ploss4 + D * Lalphanegpos(K4)*ploss5;
    pvalue4 = pvalue4 + D * Lalphanegpos(K4)*pvalue5;
    Exloss4 = Exloss4 + D * Lalphanegpos(K4)*Exloss5;
    Exvalue4 = Exvalue4 + D *
Lalphanegpos(K4)*Exvalue5;
 end
    D=1;
    if T3 <= znegpos(znegposno) , D = 0;end
    ploss3 = ploss3 + D * Lalphanegpos(K3)*ploss4;
    pvalue3 = pvalue3 + D * Lalphanegpos(K3)*pvalue4;
    Exloss3 = Exloss3 + D * Lalphanegpos(K3)*Exloss4;
    Exvalue3 = Exvalue3 + D *
Lalphanegpos(K3)*Exvalue4;
 end
    D=1;
    if T2 <= znegpos(znegposno) , D = 0;end
    ploss2 = ploss2 + D * Lalphanegpos(K2)*ploss3;
    pvalue2 = pvalue2 + D * Lalphanegpos(K2)*pvalue3;
    Exloss2 = Exloss2 + D * Lalphanegpos(K2)*Exloss3;
    Exvalue2 = Exvalue2 + D *
Lalphanegpos(K2)*Exvalue3;
 end
    D=1;
    if T1 <= znegpos(znegposno) , D = 0;end
    ploss1 = ploss1 + D * Lalphanegpos(K1)*ploss2
    pvalue1 = pvalue1 + D * Lalphanegpos(K1)*pvalue2
    Exloss1 = Exloss1 + D * Lalphanegpos(K1)*Exloss2
```

```
    Exvalue1 = Exvalue1 + D *
Lalphanegpos(K1)*Exvalue2
 end
 plossvector(t+1) = ploss1;
 pvaluevector(t+1) = pvalue1;
 Exlossvector(t+1) = Exloss1;
 Exvaluevector(t+1) = Exvalue1;
t=8
ploss1=0;
pvalue1=0;
Exloss1=0;
Exvalue1=0;
for K1 = 1: 2*Nptlevy -1
       T1 = znegpos(K1);
       ploss2=0;
       pvalue2 = 0;
       Exloss2=0;
       Exvalue2 = 0;
 for K2 = 1: 2*Nptlevy -1
       T2 = znegpos(K1)+znegpos(K2);
       ploss3=0;
       pvalue3 = 0;
       Exloss3=0;
       Exvalue3 = 0;
    for K3 = 1: 2*Nptlevy -1
       T3 = znegpos(K1)+znegpos(K2)+ znegpos(K3);
       ploss4=0;
       pvalue4 = 0;
       Exloss4=0;
       Exvalue4 = 0;
    for K4 = 1: 2*Nptlevy -1
       T4 = znegpos(K1)+znegpos(K2)+ znegpos(K3)+
znegpos(K4);
       ploss5=0;
       pvalue5 = 0;
       Exloss5=0;
       Exvalue5 = 0;
    for K5 = 1: 2*Nptlevy -1
       T5 = znegpos(K1)+znegpos(K2)+ znegpos(K3)+
znegpos(K4)+ znegpos(K5);
       ploss6=0;
       pvalue6 = 0;
       Exloss6=0;
       Exvalue6 = 0;
```

```
    for K6 = 1: 2*Nptlevy -1
       T6 = znegpos(K1)+znegpos(K2)+ znegpos(K3)+
znegpos(K4)+ znegpos(K5)+ znegpos(K6);
       ploss7=0;
       pvalue7 = 0;
       Exloss7=0;
       Exvalue7 = 0;
    for K7 = 1: 2*Nptlevy -1
       T7 = znegpos(K1)+znegpos(K2)+ znegpos(K3)+
znegpos(K4)+ znegpos(K5)+ znegpos(K6)+ znegpos(K7);
       ploss8=0;
       pvalue8 = 0;
       Exloss8=0;
       Exvalue8 = 0;
    for K8 = 1:2*Nptlevy -1
      T8 = znegpos(K1)+ znegpos(K2)+ znegpos(K3)+
znegpos(K4)+ znegpos(K5)+ znegpos(K6)+ znegpos(K7)+
znegpos(K8);
      D =1;
      if T8 <= znegpos(znegposno) , D = 0;end
      ploss8 = ploss8 + Lalphanegpos(K8)*(1-D);
      pvalue8 = pvalue8 + Lalphanegpos(K8)*((1-D)+D);
      Exloss8 = Exloss8 +
Lalphanegpos(K8)*stoplossvalue*(1-D);
      Exvalue8 = Exvalue8 +
Lalphanegpos(K8)*(stoplossvalue*(1-D)+T8*D);
    end
     D=1;
    if T7 <= znegpos(znegposno) , D = 0;end
    ploss7 = ploss7 + D * Lalphanegpos(K7)*ploss8;
    pvalue7 = pvalue7 + D * Lalphanegpos(K7)*pvalue8;
    Exloss7 = Exloss7 + D * Lalphanegpos(K7)*Exloss8;
    Exvalue7 = Exvalue7 + D *
Lalphanegpos(K7)*Exvalue8;
 end
    D=1;
    if T6 <= znegpos(znegposno) , D = 0;end
    ploss6 = ploss6 + D * Lalphanegpos(K6)*ploss7;
    pvalue6 = pvalue6 + D * Lalphanegpos(K6)*pvalue7;
    Exloss6 = Exloss6 + D * Lalphanegpos(K6)*Exloss7;
    Exvalue6 = Exvalue6 + D *
Lalphanegpos(K6)*Exvalue7;
 end
    D=1;
```

```
    if T5 <= znegpos(znegposno) , D = 0;end
    ploss5 = ploss5 + D * Lalphanegpos(K5)*ploss6;
    pvalue5 = pvalue5 + D * Lalphanegpos(K5)*pvalue6;
    Exloss5 = Exloss5 + D * Lalphanegpos(K5)*Exloss6;
    Exvalue5 = Exvalue5 + D *
Lalphanegpos(K5)*Exvalue6;
 end
    D=1;
    if T4 <= znegpos(znegposno) , D = 0;end
    ploss4 = ploss4 + D * Lalphanegpos(K4)*ploss5;
    pvalue4 = pvalue4 + D * Lalphanegpos(K4)*pvalue5;
    Exloss4 = Exloss4 + D * Lalphanegpos(K4)*Exloss5;
    Exvalue4 = Exvalue4 + D *
Lalphanegpos(K4)*Exvalue5;
 end
    D=1;
    if T3 <= znegpos(znegposno) , D = 0;end
    ploss3 = ploss3 + D * Lalphanegpos(K3)*ploss4;
    pvalue3 = pvalue3 + D * Lalphanegpos(K3)*pvalue4;
    Exloss3 = Exloss3 + D * Lalphanegpos(K3)*Exloss4;
    Exvalue3 = Exvalue3 + D *
Lalphanegpos(K3)*Exvalue4;
 end
    D=1;
    if T2 <= znegpos(znegposno) , D = 0;end
    ploss2 = ploss2 + D * Lalphanegpos(K2)*ploss3;
    pvalue2 = pvalue2 + D * Lalphanegpos(K2)*pvalue3;
    Exloss2 = Exloss2 + D * Lalphanegpos(K2)*Exloss3;
    Exvalue2 = Exvalue2 + D *
Lalphanegpos(K2)*Exvalue3;
 end
    D=1;
    if T1 <= znegpos(znegposno) , D = 0;end
    ploss1 = ploss1 + D * Lalphanegpos(K1)*ploss2;
    pvalue1 = pvalue1 + D * Lalphanegpos(K1)*pvalue2;
    Exloss1 = Exloss1 + D * Lalphanegpos(K1)*Exloss2;
    Exvalue1 = Exvalue1 + D *
Lalphanegpos(K1)*Exvalue2;
 end
 plossvector(t+1) = ploss1;% not used in calculation,
for plotting only
 pvaluevector(t+1) = pvalue1;
 Exlossvector(t+1) = Exloss1;% not used in
calculation, for plotting only
```

```
 Exvaluevector(t+1) = Exvalue1;

t=9
ploss1=0;
pvalue1=0;
Exloss1=0;
Exvalue1=0;
for K1 = 1: 2*Nptlevy -1
       T1 = znegpos(K1);
       ploss2=0;
       pvalue2 = 0;
       Exloss2=0;
       Exvalue2 = 0;
 for K2 = 1: 2*Nptlevy -1
       T2 = znegpos(K1)+znegpos(K2);
       ploss3=0;
       pvalue3 = 0;
       Exloss3=0;
       Exvalue3 = 0;
    for K3 = 1: 2*Nptlevy -1
       T3 = znegpos(K1)+znegpos(K2)+ znegpos(K3);
       ploss4=0;
       pvalue4 = 0;
       Exloss4=0;
       Exvalue4 = 0;
    for K4 = 1: 2*Nptlevy -1
       T4 = znegpos(K1)+znegpos(K2)+ znegpos(K3)+
znegpos(K4);
       ploss5=0;
       pvalue5 = 0;
       Exloss5=0;
       Exvalue5 = 0;
    for K5 = 1: 2*Nptlevy -1
       T5 = znegpos(K1)+znegpos(K2)+ znegpos(K3)+
znegpos(K4)+ znegpos(K5);
       ploss6=0;
       pvalue6 = 0;
       Exloss6=0;
       Exvalue6 = 0;
    for K6 = 1: 2*Nptlevy -1
       T6 = znegpos(K1)+znegpos(K2)+ znegpos(K3)+
znegpos(K4)+ znegpos(K5)+ znegpos(K6);
       ploss7=0;
       pvalue7 = 0;
```

```
        Exloss7=0;
        Exvalue7 = 0;
    for K7 = 1: 2*Nptlevy -1
        T7 = znegpos(K1)+znegpos(K2)+ znegpos(K3)+
znegpos(K4)+ znegpos(K5)+ znegpos(K6)+ znegpos(K7);
        ploss8=0;
        pvalue8 = 0;
        Exloss8=0;
        Exvalue8 = 0;
    for K8 = 1: 2*Nptlevy -1
        T8 = znegpos(K1)+znegpos(K2)+ znegpos(K3)+
znegpos(K4)+ znegpos(K5)+ znegpos(K6)+ znegpos(K7)+
znegpos(K8);
        ploss9=0;
        pvalue9 = 0;
        Exloss9=0;
        Exvalue9 = 0;
    for K9 = 1:2*Nptlevy -1
      T9 = znegpos(K1)+ znegpos(K2)+ znegpos(K3)+
znegpos(K4)+ znegpos(K5)+ znegpos(K6)+ znegpos(K7)+
znegpos(K8)+ znegpos(K9);
      D =1;
      if T9 <= znegpos(znegposno) , D = 0;end
      ploss9 = ploss9 + Lalphanegpos(K9)*(1-D);
      pvalue9 = pvalue9 + Lalphanegpos(K9)*((1-D)+D);
      Exloss9 = Exloss9 +
Lalphanegpos(K9)*stoplossvalue*(1-D);
      Exvalue9 = Exvalue9 +
Lalphanegpos(K9)*(stoplossvalue*(1-D)+T9*D);
    end
     D=1;
    if T8 <= znegpos(znegposno) , D = 0;end
    ploss8 = ploss8 + D * Lalphanegpos(K8)*ploss9;
    pvalue8 = pvalue8 + D * Lalphanegpos(K8)*pvalue9;
    Exloss8 = Exloss8 + D * Lalphanegpos(K8)*Exloss9;
    Exvalue8 = Exvalue8 + D *
Lalphanegpos(K8)*Exvalue9;
    end
    D=1;
    if T7 <= znegpos(znegposno) , D = 0;end
    ploss7 = ploss7 + D * Lalphanegpos(K7)*ploss8;
    pvalue7 = pvalue7 + D * Lalphanegpos(K7)*pvalue8;
    Exloss7 = Exloss7 + D * Lalphanegpos(K7)*Exloss8;
```

```
    Exvalue7 = Exvalue7 + D *
Lalphanegpos(K7)*Exvalue8;
 end
    D=1;
    if T6 <= znegpos(znegposno) , D = 0;end
    ploss6 = ploss6 + D * Lalphanegpos(K6)*ploss7;
    pvalue6 = pvalue6 + D * Lalphanegpos(K6)*pvalue7;
    Exloss6 = Exloss6 + D * Lalphanegpos(K6)*Exloss7;
    Exvalue6 = Exvalue6 + D *
Lalphanegpos(K6)*Exvalue7;
 end
    D=1;
    if T5 <= znegpos(znegposno) , D = 0;end
    ploss5 = ploss5 + D * Lalphanegpos(K5)*ploss6;
    pvalue5 = pvalue5 + D * Lalphanegpos(K5)*pvalue6;
    Exloss5 = Exloss5 + D * Lalphanegpos(K5)*Exloss6;
    Exvalue5 = Exvalue5 + D *
Lalphanegpos(K5)*Exvalue6;
 end
    D=1;
    if T4 <= znegpos(znegposno) , D = 0;end
    ploss4 = ploss4 + D * Lalphanegpos(K4)*ploss5;
    pvalue4 = pvalue4 + D * Lalphanegpos(K4)*pvalue5;
    Exloss4 = Exloss4 + D * Lalphanegpos(K4)*Exloss5;
    Exvalue4 = Exvalue4 + D *
Lalphanegpos(K4)*Exvalue5;
 end
    D=1;
    if T3 <= znegpos(znegposno) , D = 0;end
    ploss3 = ploss3 + D * Lalphanegpos(K3)*ploss4;
    pvalue3 = pvalue3 + D * Lalphanegpos(K3)*pvalue4;
    Exloss3 = Exloss3 + D * Lalphanegpos(K3)*Exloss4;
    Exvalue3 = Exvalue3 + D *
Lalphanegpos(K3)*Exvalue4;
 end
    D=1;
    if T2 <= znegpos(znegposno) , D = 0;end
    ploss2 = ploss2 + D * Lalphanegpos(K2)*ploss3;
    pvalue2 = pvalue2 + D * Lalphanegpos(K2)*pvalue3;
    Exloss2 = Exloss2 + D * Lalphanegpos(K2)*Exloss3;
    Exvalue2 = Exvalue2 + D *
Lalphanegpos(K2)*Exvalue3;
 end
    D=1;
```

```
    if T1 <= znegpos(znegposno) , D = 0;end
    ploss1 = ploss1 + D * Lalphanegpos(K1)*ploss2;
    pvalue1 = pvalue1 + D * Lalphanegpos(K1)*pvalue2;
    Exloss1 = Exloss1 + D * Lalphanegpos(K1)*Exloss2;
    Exvalue1 = Exvalue1 + D *
Lalphanegpos(K1)*Exvalue2;
 end
 plossvector(t+1) = ploss1;% not used in calculation,
for plotting only
 pvaluevector(t+1) = pvalue1;
 Exlossvector(t+1) = Exloss1;% not used in
calculation, for plotting only
 Exvaluevector(t+1) = Exvalue1;

t=10
ploss1=0;
pvalue1=0;
Exloss1=0;
Exvalue1=0;
for K1 = 1: 2*Nptlevy -1
       T1 = znegpos(K1);
       ploss2=0;
       pvalue2 = 0;
       Exloss2=0;
       Exvalue2 = 0;
 for K2 = 1: 2*Nptlevy -1
       T2 = znegpos(K1)+znegpos(K2);
       ploss3=0;
       pvalue3 = 0;
       Exloss3=0;
       Exvalue3 = 0;
    for K3 = 1: 2*Nptlevy -1
       T3 = znegpos(K1)+znegpos(K2)+ znegpos(K3);
       ploss4=0;
       pvalue4 = 0;
       Exloss4=0;
       Exvalue4 = 0;
    for K4 = 1: 2*Nptlevy -1
       T4 = znegpos(K1)+znegpos(K2)+ znegpos(K3)+
znegpos(K4);
       ploss5=0;
       pvalue5 = 0;
       Exloss5=0;
       Exvalue5 = 0;
```

```
    for K5 = 1: 2*Nptlevy -1
       T5 = znegpos(K1)+znegpos(K2)+ znegpos(K3)+
znegpos(K4)+ znegpos(K5);
       ploss6=0;
       pvalue6 = 0;
       Exloss6=0;
       Exvalue6 = 0;
    for K6 = 1: 2*Nptlevy -1
       T6 = znegpos(K1)+znegpos(K2)+ znegpos(K3)+
znegpos(K4)+ znegpos(K5)+ znegpos(K6);
       ploss7=0;
       pvalue7 = 0;
       Exloss7=0;
       Exvalue7 = 0;
    for K7 = 1: 2*Nptlevy -1
       T7 = znegpos(K1)+znegpos(K2)+ znegpos(K3)+
znegpos(K4)+ znegpos(K5)+ znegpos(K6)+ znegpos(K7);
       ploss8=0;
       pvalue8 = 0;
       Exloss8=0;
       Exvalue8 = 0;
    for K8 = 1: 2*Nptlevy -1
       T8 = znegpos(K1)+znegpos(K2)+ znegpos(K3)+
znegpos(K4)+ znegpos(K5)+ znegpos(K6)+ znegpos(K7)+
znegpos(K8);
       ploss9=0;
       pvalue9 = 0;
       Exloss9=0;
       Exvalue9 = 0;
    for K9 = 1: 2*Nptlevy -1
       T9 = znegpos(K1)+znegpos(K2)+ znegpos(K3)+
znegpos(K4)+ znegpos(K5)+ znegpos(K6)+ znegpos(K7)+
znegpos(K8)+ znegpos(K9);
       ploss10=0;
       pvalue10 = 0;
       Exloss10=0;
       Exvalue10 = 0;
   for K10 = 1:2*Nptlevy -1
      T10 = znegpos(K1)+ znegpos(K2)+ znegpos(K3)+
znegpos(K4)+ znegpos(K5)+ znegpos(K6)+ znegpos(K7)+
znegpos(K8)+ znegpos(K9)+ znegpos(K10);
      D =1;
      if T10 <= znegpos(znegposno) , D = 0;end
      ploss10 = ploss10 + Lalphanegpos(K10)*(1-D);
```

```
      pvalue10 = pvalue10 + Lalphanegpos(K10)*((1-
D)+D);
      Exloss10 = Exloss10 +
Lalphanegpos(K10)*stoplossvalue*(1-D);
      Exvalue10 = Exvalue10 +
Lalphanegpos(K10)*(stoplossvalue*(1-D)+T10*D);
    end
    D=1;
    if T9 <= znegpos(znegposno) , D = 0;end
    ploss9 = ploss9 + D * Lalphanegpos(K9)*ploss10;
    pvalue9 = pvalue9 + D *
Lalphanegpos(K9)*pvalue10;
    Exloss9 = Exloss9 + D *
Lalphanegpos(K9)*Exloss10;
    Exvalue9 = Exvalue9 + D *
Lalphanegpos(K9)*Exvalue10;
    end
    D=1;
    if T8 <= znegpos(znegposno) , D = 0;end
    ploss8 = ploss8 + D * Lalphanegpos(K8)*ploss9;
    pvalue8 = pvalue8 + D * Lalphanegpos(K8)*pvalue9;
    Exloss8 = Exloss8 + D * Lalphanegpos(K8)*Exloss9;
    Exvalue8 = Exvalue8 + D *
Lalphanegpos(K8)*Exvalue9;
    end
    D=1;
    if T7 <= znegpos(znegposno) , D = 0;end
    ploss7 = ploss7 + D * Lalphanegpos(K7)*ploss8;
    pvalue7 = pvalue7 + D * Lalphanegpos(K7)*pvalue8;
    Exloss7 = Exloss7 + D * Lalphanegpos(K7)*Exloss8;
    Exvalue7 = Exvalue7 + D *
Lalphanegpos(K7)*Exvalue8;
 end
    D=1;
    if T6 <= znegpos(znegposno) , D = 0;end
    ploss6 = ploss6 + D * Lalphanegpos(K6)*ploss7;
    pvalue6 = pvalue6 + D * Lalphanegpos(K6)*pvalue7;
    Exloss6 = Exloss6 + D * Lalphanegpos(K6)*Exloss7;
    Exvalue6 = Exvalue6 + D *
Lalphanegpos(K6)*Exvalue7;
 end
    D=1;
    if T5 <= znegpos(znegposno) , D = 0;end
    ploss5 = ploss5 + D * Lalphanegpos(K5)*ploss6;
```

```
    pvalue5 = pvalue5 + D * Lalphanegpos(K5)*pvalue6;
    Exloss5 = Exloss5 + D * Lalphanegpos(K5)*Exloss6;
    Exvalue5 = Exvalue5 + D *
Lalphanegpos(K5)*Exvalue6;
 end
    D=1;
    if T4 <= znegpos(znegposno) , D = 0;end
    ploss4 = ploss4 + D * Lalphanegpos(K4)*ploss5;
    pvalue4 = pvalue4 + D * Lalphanegpos(K4)*pvalue5;
    Exloss4 = Exloss4 + D * Lalphanegpos(K4)*Exloss5;
    Exvalue4 = Exvalue4 + D *
Lalphanegpos(K4)*Exvalue5;
 end
    D=1;
    if T3 <= znegpos(znegposno) , D = 0;end
    ploss3 = ploss3 + D * Lalphanegpos(K3)*ploss4;
    pvalue3 = pvalue3 + D * Lalphanegpos(K3)*pvalue4;
    Exloss3 = Exloss3 + D * Lalphanegpos(K3)*Exloss4;
    Exvalue3 = Exvalue3 + D *
Lalphanegpos(K3)*Exvalue4;
 end
    D=1;
    if T2 <= znegpos(znegposno) , D = 0;end
    ploss2 = ploss2 + D * Lalphanegpos(K2)*ploss3;
    pvalue2 = pvalue2 + D * Lalphanegpos(K2)*pvalue3;
    Exloss2 = Exloss2 + D * Lalphanegpos(K2)*Exloss3;
    Exvalue2 = Exvalue2 + D *
Lalphanegpos(K2)*Exvalue3;
 end
    D=1;
    if T1 <= znegpos(znegposno) , D = 0;end
    ploss1 = ploss1 + D * Lalphanegpos(K1)*ploss2;
    pvalue1 = pvalue1 + D * Lalphanegpos(K1)*pvalue2;
    Exloss1 = Exloss1 + D * Lalphanegpos(K1)*Exloss2;
    Exvalue1 = Exvalue1 + D *
Lalphanegpos(K1)*Exvalue2;
 end
 plossvector(t+1) = ploss1;% not used in calculation,
for plotting only
 pvaluevector(t+1) = pvalue1;
 Exlossvector(t+1) = Exloss1;% not used in
calculation, for plotting only
 Exvaluevector(t+1) = Exvalue1;
```

```
for I =1:t+1
   time(I) = I-1;
end
figure(4)
plot(time,Exvaluevector,'k*-')
xlabel(' t ')
ylabel('Expected value at t, with fixed stop-loss')
title('S & P 500')
Totalp(2)=1;% As Lalphanegpos is normalized.  The
total probability is 1 at t=1
Totalex(2)=Exvaluevector(2);
plosssum=0;
Exlosssum=0;
for I=3:t+1
   plosssum = plosssum + plossvector(I-1);% add up
prob of stopping out in previous days, 35M142
   Exlosssum = Exlosssum+Exlossvector(I-1);% add up
expected values of stopping out for previous days
   Totalp(I)=plosssum+pvaluevector(I);% Totalp is the
sum of the probability of previous days losses and
last day's cshing out
   Totalex(I)=Exlosssum+Exvaluevector(I);%Totalex is
the expected value of previous days losses and last
day's gain
end
   Expectedt(2)=1;%Expectedt is the average number of
t units of the trade when the trader chooses to cash
out on the t th unit
for I=3:t+1
   Sum=0;
   for J=2:I-1
      Sum =Sum + plossvector(J)*(J-1);
   end
   Expectedt(I)= Sum +pvaluevector(I)*(I-1);
end
for I=2:t+1
   TotalexDt(I)=Totalex(I)/Expectedt(I);
end
figure(5)
plot(time(2:t+1),Exlossvector(2:t+1),'k*-')% last
point of Exlossvector actually not used in
calculation.  It is calculated for plotting purpose
only.
```

```
xlabel(' t ')
ylabel('Expected loss at time t, with fixed stop loss
')
title('S & P 500')
figure(6)
plot(time(2:t+1),Totalex(2:t+1),'k*-')
xlabel(' t ')
ylabel('Total expected value, with fixed stop-loss')
title('S & P 500')
figure(7)
plot(time(2:t+1),plossvector(2:t+1),'k*-')
xlabel(' t ')
ylabel('plossvector')
title('S & P 500')
figure(8)
plot(time(2:t+1),pvaluevector(2:t+1),'k*-')
xlabel(' t ')
ylabel('pvaluevector, with fixed stop-loss')
title('S & P 500')
figure(9)
plot(time(2:t+1),Totalp(2:t+1),'k*-')
xlabel(' t ')
ylabel('Totalp, with fixed stop-loss')
title('S & P 500')
figure(10)
plot(time(2:t+1),Expectedt(2:t+1),'k*-')
xlabel(' t ')
ylabel('Average time, with fixed stop-loss')
title('S & P 500')
figure(11)
plot(time(2:t+1),TotalexDt(2:t+1),'k*-')
xlabel(' t ')
ylabel('Total expected value/Average time , with
fixed stop-loss ')
title('S & P 500')
```

Bibliography

Anderson, D. R., Sweeney, D. J. and Thomas, A. W., *Statistics for Business and Economics 9e*, Thompson Southwestern (2005).

Antoniou, I., Ivanov, Vi. V., Ivanov, Va. V. and Zrelov, P. V., "On the log-normal distribution of stock market data", *Physica A*, V331, p617 – 638 (2004).

Appel, Gerald, *Winning Market Systems: 83 Ways to Beat the Market*, Traders Press (1991).

Bernstein, J., *Cycles of Profit*, HarperBusiness (1991).

Bevington, P. R., *Data Reduction and Error Analysis for the Physical Sciences*, McGraw-Hill (1969).

Brigham, E. O., *The Fast Fourier Transform*, Prentice-Hall (1974).

Broesch, J. D., *Digital Signal Processing Demystified*, LLH Technology Publishing (1997).

Burrus, C. S., Gopinath, R. A. and Guo, H., *Introduction to wavelets and wavelet transforms, A primer*, Prentice-Hall (1998).

Butkov, E., *Mathematical Physics*, Addison-Wesley (1968).

Casti, J. L., *Complexification*, Harper Perennial (1995).

Crow, E. L. and Shimizu, K., *Lognormal Distributions : Theory and Applications*, Marcel Dekker, Inc. (1988).

Daubechies, I., Ten Lectures on Wavelets, Siam (1992).

Ehlers, J. F., *MESA and Trading Market Cycles*, John Wiley & Sons (1992).

Ehlers, J. F., *Rocket Science for Traders*, John Wiley & Sons (2001).

Elder, A., *Trading for a Living*, John Wiley & Sons (1993).

Elder, A., *Come into my trading room*, John Wiley & Sons (2002).

Freund, J. E.,*Mathematical Statistics*, Prentice Hall (1992).

Hamming, R. W., *Digital Filters*, Third Edition, Dover Publications (1989).

Hanselman, D. and Littlefield, B., *The Student Edition of MATLAB*, Version 5 User's Guide, Prentice Hall (1997).

Hayes, M. H., *Digital Signal Processing, Schaum's Outlines*, McGraw-Hill (1999).

Huang, K., *Statistical Mechanics*, John Wiley & Sons, Inc. (1963).

Hubbard, B. B., *The World according to Wavelets*, Second Edition, A. K. Peters, Ltd.(1998).

Johnson, N. F., Jeffries, P. and Hui P. M., *Financial Market complexity*, Oxford University Press (2003).

Kaiser, G., *A Friendly Guide to Wavelets*, Birkhauser (1994).

Kaplan, W., *Advanced Calculus*, Addison-Wesley (1959).

Lillo, F. and Mantegna, R. N., "Power-law relaxation in a complex system: Omori law after a financial market crash", *Physical Review E*, V68, p016119-1 – 016119-5 (2003).

Lo A. W. and MacKinlay, A. C., *A Non-Random Walk down Wall Street*, Princeton University Press (1999).

Lyons, R. G., *Understanding Digital Signal Processing*, Addison-Wesley (1997).

Mak, D. K., *Science of Financial Market Trading*, World Scientific (2003).

Mallat, S., *A Wavelet Tour of Signal Processing*, Second Edition, Academic Press (1999).

Mandelbrot, B. B., *The fractal geometry of nature*, W. H. Freeman and Company (1983).

Mandelbrot, B. B., *Fractals and scaling in Finance*, Springer (1997).

Mantegna, R. N. and Stanley, H. E., "Scaling behaviour in the dynamics of an economic index", *Nature*, V376, p46-49 (1995).

Mantegna, R. N. and Stanley, H. E., *An Introduction to Econophysics, Correlations and Complexity in Finance*, Cambridge University Press (2000).

Meyer, P. L., *Introductory Probability and Statistical Applications*, Addison-Wesley (1965).

Michael F. and Johnson, M. D., "Financial market dynamics", *Physica A*, V320, p525-534 (2002).

Oppenheim, A. V., Schafer, R. W., with Buck, J. R., Discrete-Time Signal Processing, Second Edition, Prentice-Hall (1999).

Prechter Jr., R. R. and Frost, A. J., *Elliott Wave Principle*, New Classic Library (1990).

Pring, M. J., *Technical Analysis Explained, The successful investor's guide to spotting investment trends and turning points*, Third Edition, McGraw-Hill Inc. (1991).

Proakis, J. G. and Manolakis, D. G., *Digital Signal Processing, Principles, Algorithms and Applications*, Third Edition, Prentice-Hall (1996).

Protter, M. H. and Morrey, C. B., Jr., *Calculus with Analytic Geometry*, Addison-Wesley (1963).

Rao, R. M. and Bopardikar, A. S., *Wavelet Transforms, Introduction to Theory and Applications*, Addison-Wesley (1998).

Sornette, D., *Why Stock Markets Crash*, Princeton University Press (2003).

Stanley, E. H., *Introduction to Phase Transitions and Critical Phenomena*, Oxford University Press (1971).

Strang, G. and Nguyen, T., *Wavelets and Filter Banks*, Wellesley-Cambridge Press, Revised Edition) (1997).

Tsallis C. and Bukman, "Anomalous diffusion in the presence of external forces: Exact time-dependent solutions and their thermostatistical basis", *Phys. Rev. E*, V54, R2197 – R2200 (1996).

Waldrop, M. M., *Complexity*, Simon & Schuster (1992).

Index